Progress in Molecular and Subcellular Biology 8

8 Progress in Molecular and Subcellular Biology

With Contributions by
J. R. Bamburg, T. L. German, F. E. Hahn
G. Knudson, R. F. Marsh, J. J. O'Neill

Edited by F. E. Hahn, D. J. Kopecko
W. E. G. Müller

Managing Editor: F. E. Hahn

Springer-Verlag
Berlin Heidelberg New York Tokyo 1983

Professor Fred E. Hahn, Ph. D.
Division of Communicable Diseases and Immunology
Walter Reed Army Institute of Research
Washington, D.C. 20307, USA

With 36 Figures

ISBN-13: 978-3-642-69230-7 e-ISBN-13: 978-3-642-69228-4
DOI: 10.1007/978-3-642-69228-4

Library of Congress Catalog Card Number 75-79748

Softcover reprint of the hardcover 1st edition 1983

Offsetprinting and Bookbinding: Konrad Triltsch, Graphischer Betrieb, 8700 Würzburg.
2131/3130-543210

Contents

T.L. GERMAN and R.F. MARSH: The Scrapie Agent: A Unique Self-Replicating Pathogen (With 4 Figures)

J.J. O'NEILL: Non-Cholinesterase Effects of Anticholinesterases (With 12 Figures)

Contributors

BAMBURG, J.R., Chairman, Department of Biochemistry, Colorado State University, Ft. Collins, CO 80523, USA

GERMAN, T.L., Department of Veterinary Science, 1655 Linden Drive, University of Wisconsin, Madison, WI 53706, USA

HAHN, F.E., Walter Reed Army Institute of Research, Division of Communicable Diseases and Immunology, Washington, D.C. 20307, USA

KNUDSON, G., U.S. Army Medical Research Institute of Infectious Diseases, Ft. Detrick, Frederick, MD 21701, USA

MARSH, R.F., Department of Veterinary Science, 1655 Linden Drive, University of Wisconsin, Madison, WI 53706, USA

O'NEILL, J.J., Chairman, Pharmacology Department, Temple University, 3420 N. Broad Street, Philadelphia, PA 19140, USA

Penicillin Until 1957 *

Fred E. Hahn

A. Introduction

It may seem paradoxical to include in a progress volume a contribution whose title suggests that it is retrospective and deals with the first 30 years of research on the mode and mechanism of action of penicillin. To the surprise of the author, his studies into the earlier history of this research field have brought to light a wealth of observations and experimental findings which are forgotten and no longer read. Moreover, some of this material has a distinctly contemporary ring to it.

How can this be the case? Scientists are brought up with a view of the aggregative accumulation of scientific knowledge, resembling the building of a house in which brick is mortared upon brick, each earlier structural component carrying the subsequent accretion. But in reality, things do not appear to be so simple.

Figure 1, taken from a book entitled *The Growth of Knowledge* (Kochen, ed. 1967), has been assembled by De Solla Price from data of Garfield. It depicts for each year between 1860 and 1960, the ratios of the number of citations in 1961 to the number of articles published in each year. It illustrates that during the first 20 years after publication, the bibliographical use of articles declines steeply and systematically and then continues to decline more gradually until it approaches statistically a ratio of one citation of each paper per year.

There is an "immediacy factor" in the use of published knowledge which means that about 30 per cent of all references are to the recent research front while every year about 10 per cent of all publications "die", not to be cited and reviewed again. It means that if recent work is not cited rather quickly, it may not be cited and reviewed at all, but simply be buried in the growing archive of scientific literature.

This will not only occur with articles of mediocre quality, but also with those which, in certain respects, are so far advanced that the field perceives them as "unzeitgemäss" or premature. In fact, Stent (1972) wrote an interesting study, entitled *Prematurity and uniqueness in scientific discovery*.

---

* The views of the author do not purport to reflect the position of the Department of the Army or the Department of Defense

Progress in Molecular and Subcellular Biology, Vol. 8, edited by F.E. Hahn

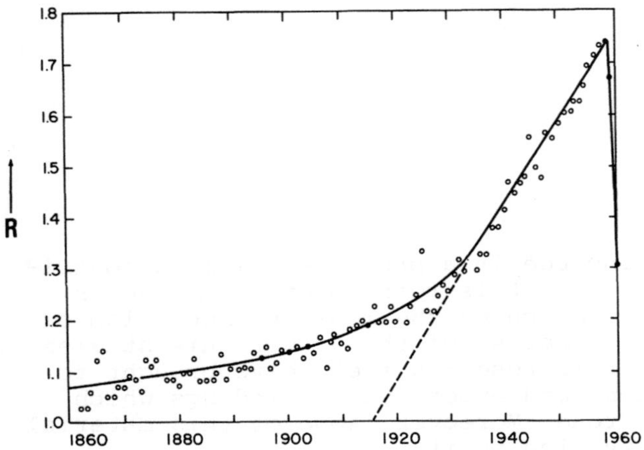

Fig. 1. Ratios of the number of 1961 citations to the number of papers published in each of the years 1860 through 1960 (De Solla Price in Kochen, 1967)

During a tenure at the University of Heidelberg, I became acquainted with the Nobel Laureate Richard Kuhn and learned that he was spending a considerable portion of his time reading the older chemical literature. As a direct result of his studies, he introduced column chromatography in 1932 which had been published 28 years earlier by Tswett and completely missed by the field. Kuhn also introduced in 1941 triphenyltetrazolium chloride as an irreversible reduction indicator, going back to an original publication of von Pechmann and Runge in 1894. The discovery of cytochromes by MacMunn in 1883 and their rediscovery by Keilin in 1925, as well as the classical genetic paper of Gregor Mendel of 1865 which was confirmed as late as 1900 by de Vries are additional examples of important articles in the source literature being overlooked and promptly forgotten.

B. The Discovery of Penicillin

For a while, this nearly happened with penicillin. The antibiotic was discovered by Fleming in 1928, although his published report dates from May 1929. I shall quote the introduction to his report (Fleming, 1929).

"While working with staphylococcus variants, a number of culture plates were set aside on the laboratory bench and examined from time to time. In these examinations, these plates were necessarily exposed to the air and they became contaminated with various microorganisms. It was noticed that around a large colony of a contaminating mould, the staphylococcus colonies became transparent and were obviously undergoing lysis."

"Subcultures of this mould were made and experiments conducted with a view to ascertaining something of the properties of the bacteriolytic substance which had evidently been formed in the mould culture and which had diffused into the surrounding medium. It was found that broth in which the mould had been grown at room temperature for one or two weeks had acquired marked inhibitory, bactericidal and bacteriolytic properties to many of the more common pathogenic bacteria."

Fleming carried out a series of mostly bacteriological studies with the fermentation broth of the fungus. And summarizes some of them as follows:

"The active agent is readily filterable and the name penicillin has been given to filtrates of the broth culture of the mould."

"The action is very marked on the pyogenic cocci and the diphtheria group of bacilli. Many bacteria are quite insensitive, e.g. the coli-typhoid group, the influenza-bacillus group, and the enterococcus."

"Penicillin is non-toxic to animals in enormous doses and is non-irritant. It does not interfere with leucocytic function to a greater degree than does ordinary broth. It is suggested that it may be an effective antiseptic for application to, or injection into, areas infected with penicillin-sensitive microbes."

The entire paper on the discovery of penicillin does not once contain the term chemotherapy. Fleming was interested throughout his medical life in the natural antibacterial action of blood and antiseptics and makes a special point of mentioning that in vitro penicillin, which completely inhibited the growth of staphylococci in a dilution of 1 in 600, did not interfere with leucocyte function to a greater extent than does ordinary broth. By relating the action of penicillin to phagocytosis and restricting his experimentation almost exclusively to in vitro antibacterial testing he failed to call the attention of the field to the potential discovery of an antibacterial drug for systemic administration. Whereas Fleming's discovery was undoubtedly widely discussed, it was cited in the subsequent ten years no more than three times.

In 1932, three years after the discovery was published, Clutterbuck, Lovell and Raistrick published a paper which was No. 127 in a series entitled *Studies on the Biochemistry of Microorganisms*. They succeeded in fermenting penicillin from Fleming's strain in a simple mineral medium with glucose as the source of carbon, in the hope of ready isolation. But the antibacterial potency was lost during evaporation of an ether solution in a stream of air or by evaporation in vacuo at 40-45^{0} in acid and alkaline solutions.

Three years later, in 1935, Roger Reid of Pennsylvania State College repeated the fermentation in synthetic medium. He found the antibacterial activity relatively thermo-stable but could not separate it from the rest of the filtrate by dialysis, absorption on charcoal, or distillation at low temperature.

Five years later (1940), in New York, Siegbert Bornstein used the filtrate of the penicillium cultures of Fleming's organism in studies on bacterial taxonomy. From the history of the first 10 years, it appears that the field did not appreciate the significance of Fleming's discovery and did not rally to an intense and systematic effort of isolating penicillin, determining its structure, and trying it as a chemotherapeutic drug.

It is of interest to ask why Fleming's own work on his discovery ceased despite his appreciation of the fact that as an antiseptic, penicillin broth differed so drastically from other known antiseptics. In 1940 he wrote "We have been using it in the laboratory for over 10 years as a method of differential culture. It was used in a few cases as a local antiseptic but although it gave reasonably good results, the trouble of making it seemed not worthwhile," and one year later: "a few tentative observations had been made on the local application of the unconcentrated culture to septic wounds. Although the results were considered favourable, ... it was not considered that the production of penicillin for the treatment of these was practicable, owing to the lability of the active principle in solution." And in 1945, after penicillin as a systemic drug had become a reality, Fleming wrote: "When I saw certain changes on my culture plates as the result of the mould contaminant, I had not the slightest suspicion that I was at the beginning of something extraordinary." The three preceding quotations of Fleming are cited by Florey et al. (1949). Thus from the history of the first 12 years, it appears that the discovery of penicillin was at risk of being forgotten, like the other scientific discoveries mentioned in the introduction. This is even more surprising since sulfonamides had been introduced in 1935 and showed that substances did exist which could control systemic bacterial infections after absorption into the blood stream.

C. Penicillin as a Chemotherapeutic Drug

The first human cases to be cured by penicillin were four babies and a colliery manager. The work was carried out by Paine in 1931 but was never published in the medical literature. Only after penicillin had become famous, were these first cures discovered by investigating journalists and authors (Wilson, 1976). Paine produced his own penicillin broth. Four infants in Sheffield, two with staphylococcal and two with gonococcal eye infections were treated by infusion of penicillin broth into the eyes every 4 hours. After three days, both gonococcal infections and one of the staphylococcal infections were cured.

The colliery manager had one eye penetrated by a small chip of rock when he was down in the mine. He developed a serious eye infection with *Pneumococcus* which rendered impossible the surgical removal of the stone sliver. Fourty-eight hours of continuous irrigation of the eye with penicillin broth cured the infection, the stone chip was removed routinely, and the patient's eyesight was saved.

Dr. Paine did not continue his work with penicillin. One wonders how the history of penicillin would have developed, if these clinical results has been published. But the work went unnoticed.

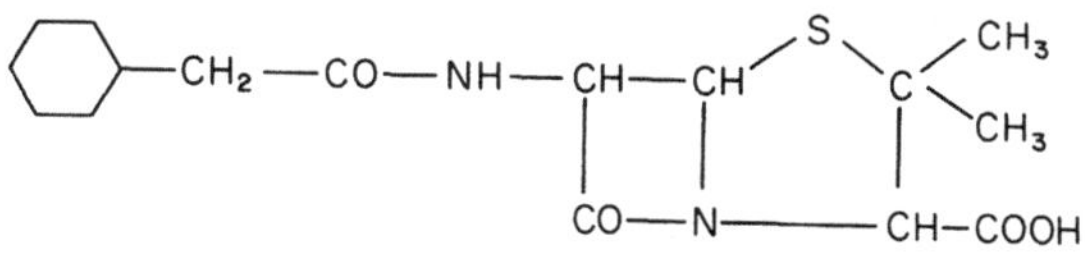

Fig. 2. Structure of Penicillin G

The decisive breakthrough came in 1940 through the work of Chain, Florey and other scientists at the Sir William Dunn School of Pathology in Oxford (Chain et al. 1940). These investigators obtained penicillin as an impure brown powder and gave their first publication the title, *Penicillin as a Chemotherapeutic Agent*. They ascertained the non-toxicity of penicillin solutions injected into laboratory rodents and showed that the drug was active in vivo against three Gram-positive organisms against which it had shown activity in vitro. One year later there followed a long paper (Abraham et al. 1941) on a therapeutic trial involving 10 cases of human infections: five patients with staphylococcal and streptococcal infections were treated intravenously, a baby with a persistant staphylococal urinary infection, by mouth and four cases of eye infections by local application. In all cases, a favorable therapeutic response was obtained.

In the same year, 1941, a paper was read by Dawson, Gladys Hobby, Karl Meyer and Chaffee before the Annual Meeting of the American Society for Clinical Investigation, entitled *Penicillin as a Chemotherapeutic Agent* in which inter alia it was reported that penicillin protected mice against 100,000,000 lethal doses of hemolytic streptococci. Hence, in 1941, Penicillin had been put on the map of chemotherapeutic drugs.

Despite these early successes in characterizing penicillin as an extraordinarily potent chemotherapeutic drug, two important items of scientific information were still missing. The drug had not been purified from fermentation mixtures, and its chemical structure had not been determined. By 1943, the recognition of the potential medical importance of penicillin resulted in the restriction of information on its chemical nature. Although some 40 chemical laboratories in the United Kingdom and the United States worked on penicillin, their results were exchanged and communicated only in the form of internal reports, and the first brief summaries of the results appeared in print in Nature (1945) and in Science (1945) after the cessation of hostilities.

D. Early Studies on Mode of Action

While the confidential work on the purification, structure, and derivatization of penicillin was under way, a first study on the mechanism of action of the crude antibiotic was published in 1942 by Hobby, Meyer and Chaffee. Penicillin was found to be bactericidal for Gram-positive cocci, and the rate of killing of the bacteria was of first order with time. There was a limit beyond which an increase in the concentration of penicillin did not accelerate the rate of killing. The authors did not observe bacterial lysis, and the amount of penicillin in 48 h cultural filtrates was undiminished.

Most important, penicillin only killed bacteria under conditions which permitted the growth of control cultures. This observation that active bacterial growth was required for the bactericidal action of the antibiotic was made repeatedly in subsequent studies. Today, the interpretation of this finding would be that the drug gives rise to some form of unbalanced and lethal biosynthesis.

E. Paradoxical Inhibitory Zone Phenomena

When a well containing penicillin solution or a paper disc impregnated with penicillin are placed on a culture plate, seeded, for example, with *Staphylococcus aureus*, the zone phenomenon, depicted in Fig. 3 is typically observed (Pratt and Dufrenoy 1949). Close to the source of penicillin is a zone of growth inhibition, the diameter of which is, within limits, a function of the concentration of penicillin under test. The larger part of the plate exhibits normal bacterial growth. At the boundary of the zone of inhibition, however, a ring of enhanced growth of the bacterial population is to be seen. This phenomenon is reproduced easily and suggests the existence of a critical threshold of penicillin concentration below which it is not growth E_{max} inhibitory but, in fact, stimulates bacterial growth.

Staining of these culture plates with redox indicators revealed that the rings of enhanced growth exhibited very strong reductive activity, but the subsequent literature has failed to yield a biochemical explanation of the paradoxical zone phenomenon.

Penicillin is not the only growth inhibitor which gives rise to paradoxical zones of growth stimulation. The same phenomenon has

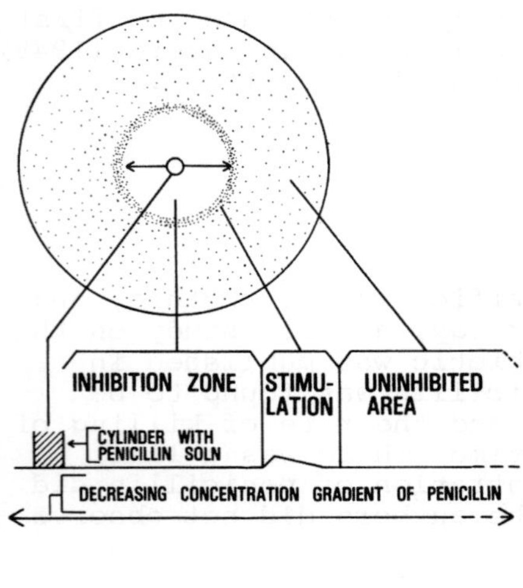

Fig. 3. Diagrammatic representation of a penicillin assay plate, showing ring of enhanced growth of *Staphylococcus aureus* on uniformly seeded culture plates supplied with a cyclinder of penicilling solution: surface view above, sectional view below. (Pratt and Dufrenoy 1949)

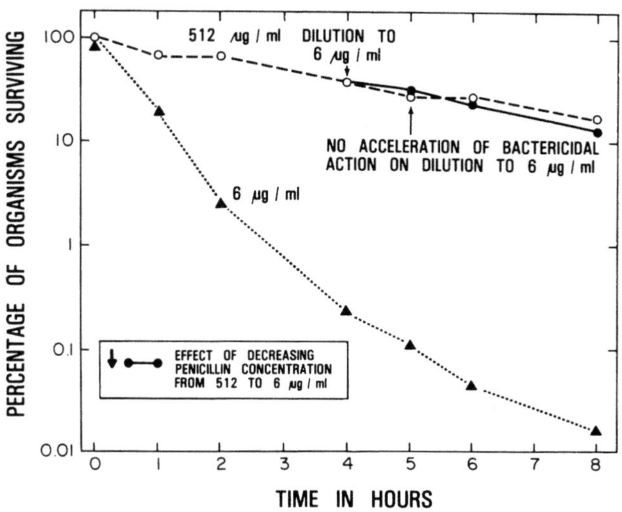

Fig. 4. The persistent effect of brief exposure to high concentrations of penicillin on the death rate of *Streptococcus faecalis* (Eagle 1951)

been demonstrated for sulfanilamide and for mercuric chloride (Lamanna and Shapiro 1943).

An obverse zone phenomenon in the bactericidal action of penicillin has been demonstrated in liquid culture (Eagle and Musselman 1948; Eagle 1951). It was shown that certain strains of bacteria are killed more rapidly by low concentrations of penicillin than by high concentrations. When such zone-reacting bacteria are first exposed to optimal, i.e. low concentrations of penicillin at which they die rapidly and subsequently high concentrations of penicillin are supplied to the cultures, the rate of death is immediately retarded to that characteristic for higher concentrations.

The paradoxical slowing down of the bactericidal effect by high concentrations of penicillin persists after the removal of the drug (Eagle 1951). If zone-sensitive bacteria are first exposed to high concentrations of penicillin at which they die at a slow rate, and the mixture is diluted after several hours incubation to the maximally effective low level of penicillin, the rate of death is not accelerated to that characteristic for the lower penicillin concentration, but the organisms continue to die at the slow rate initially established by the high concentration. Exposure to high concentrations for as short as 15 minutes, i.e. before an appreciable number of bacteria has been killed, suffices for the subsequent protection against rapid killing by low penicillin concentrations.

F. Morphological Changes in Bacteria

Beginning with the original observation of Gardner (1940), a considerable literature (reviewed by Florey et al. 1949) described the bizarre morphological changes, caused by the action of penicillin, frequently at sub-inhibitory concentrations as small as one tenth the amount required for complete growth inhibition. Large forms of irregular shape were observed in Gram-positive and Gram-negative species. Cocci produced swollen forms and bacilli formed long filaments. These malformations were attributed by Gardner to an interference with the fission of multiplying cells.

Morphological changes caused by penicillin included the formation of L-forms (Bringmann 1952; Lederberg 1956). Such penicillin-induced L-forms can revert to normal morphology when cultivated in the absence of the antibiotic (Johnstone et al. 1950; Lederberg and St. Clair 1958).

Perhaps the most important article on penicillin-induced morphological changes was published by Duguid in 1946 which remained unnoticed for 10 years. A series of sensitive and relatively resistant bacteria was grown on blocks of nutrient agar which incorporated different concentrations of penicillin and which were mounted between a slide and coverslip under an incubated microscope.

Figure 5 shows the effect of different concentrations of penicillin on the growth of *E. coli* observed over a graded period of time. "Up to the stage of filament formation and swelling, the abnormal cells were apparently alive, for growth had been proceeding and normal motility was exhibited in the case of the motile strains. Lysis, and thus death, of the filamentous cell was in most cases initiated by the gradual or sudden protrusion of one or sometimes more bubbles of protoplasm; following this, the filament became pale or even disappeared entirely. Some filamentous cells underwent lysis without any visable protoplasmic protrusion, and some without even having developed a swelling."

"The morphological changes described above as failure of proper cell division and the ready occurrence of swelling and protoplasmic protrusion, suggest that penicillin in these concentrations interferes specifically with the formation of the outer supporting cell wall, while otherwise allowing growth to proceed until the organism finally bursts its defective envelope and so undergoes lysis. In the higher concentrations, penicillin must act somewhat differently."

The significance of this 1946 publication is that it postulated on purely morphological grounds the theory of penicillin as an inhibitor of bacterial cell-wall biosynthesis which was suggested on biochemical grounds in 1957 by Park and Strominger and dominated the thinking about the mechanism of action of penicillin in subsequent years.

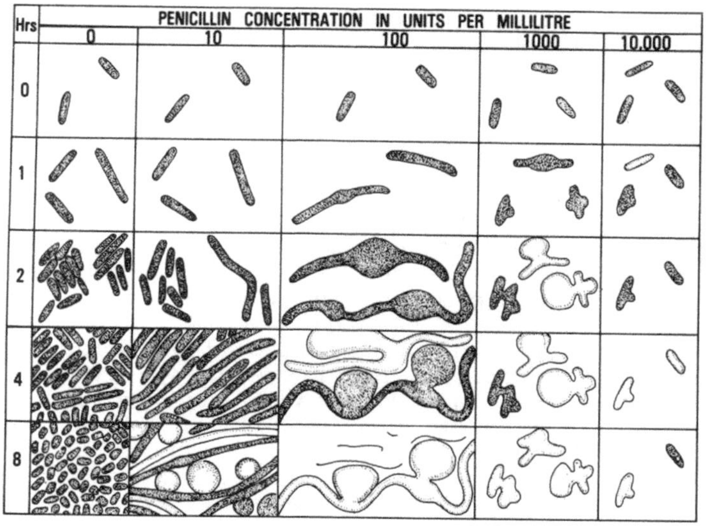

Fig. 5. Morphological effects of penicillin on growing *Escherichia coli* (Duguid 1946)

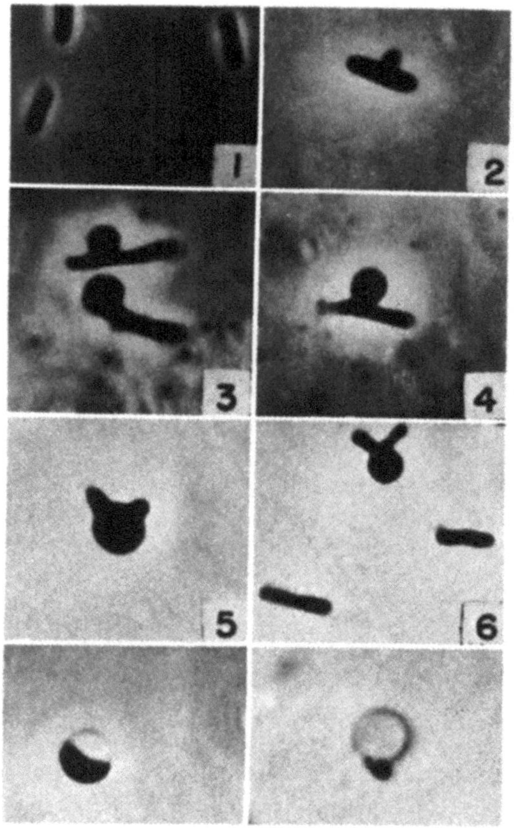

Fig. 6. Morphological effects of penicillin on *Escherichia coli*, growing in liquid medium with 0.48 M sucrose for osmotic protection (Hahn and Ciak 1957)

In 1956 and 1957, several groups of authors studied penicillin-induced changes in bacterial morphology in liquid cultures. Liebermeister and Kellenberger (1956) worked with *Proteus vulgaris* and obtained a systematic transition of bacillary into globular forms, especially when penicillin was added at the end of the exponential phase of growth. Earlier addition of penicillin produced lysis of the these cultures.

Lederberg (1956) and Hahn and Clark (1957) studied *E. coli* cultures to which sucrose had been added for osmotic protection. Figure 6 shows the sequence of morphological changes of *E. coli* B photographed under the phase-contrast microscope in my laboratory. The bacterial rods produced central or terminal globular extrusions that increased in size while the bacterial cell walls became correspondingly empty of cytoplasm. Later the globes either separated from the cell walls or retained parts of them attached, giving a typical rabbit-ear appearance. Finally, the globular structures underwent partial vacuolization, showing many crescent-shaped forms. Eventually, they released their entire content, leaving as formed elements only circular "ghosts" that probably represented empty cytoplasmic membranes."

G. Bacterial Lysis by Penicillin

While the original discovery of Fleming concerned the lysis of fully grown cultures of *Staphylococci* by penicillin, elaborated by a mold culture, the years 1943-1946 saw the publication of a rather extensive literature on the progress of lysis in liquid cultures of *Staphylococcus* which was followed turbidimetrically.

Figure 7 from a paper of Chain and Duthi (1945) shows the typical result of this experimental effort. There was general agreement that the turbidity of a young culture in nutrient medium, containing penicillin, first increased and then gradually decreased until bacterial lysis was complete. The initial increase in turbidity was alternately interpreted as being due to multiplication of the bacteria or as the result of swelling of staphylococci before lysis. Chain and Duthie compared their turbidimetric measurements with the total cell counts and showed that there was no increase in the number of cells.

In 1957 after a hiatus of more than 10 years, Hahn and Ciak and Prestidge and Pardee published results on the penicillin E_{max} induced lysis of *Escherichia coli*. The first two authors correlated the morphological destruction of the bacteria with turbidimetric measurements of lysis.

In the absence of sucrose for osmotic protection, turbidity of liquid cultures slightly increased during the first hour of penicillin action and then rapidly decreased. Aerated cultures began to foam, and masses of macroscopic long strands appeared that gave the impression of highly polymerized material. Perchloric acid extracts of such collected strands had absorption spectra resembling those of nucleic acids and contained quantities of

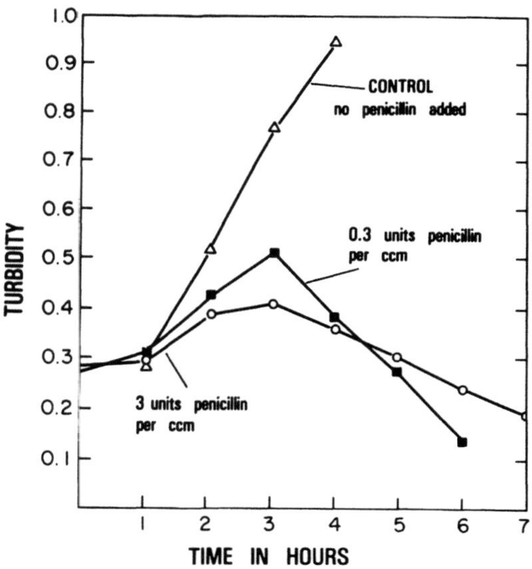

Fig. 7. Penicillin-induced lysis of *Staphylococcus*, growing in liquid medium (Chain and Duthi 1945)

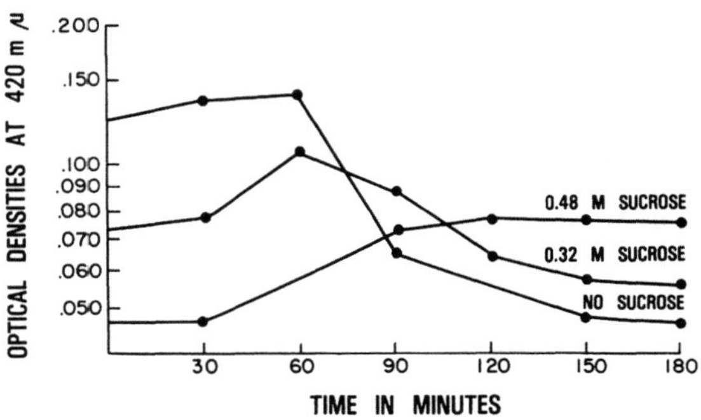

Fig. 8. Penicillin-induced lysis of *Escherichia coli* growing in liquid media with and without sucrose for osmotic protection (Hahn and Ciak 1957)

pentose and deoxypentose which suggested the presence of RNA and DNA in a ratio of 3.5 to 1.

Somewhat slower lysis occurred in the presence of 0.32 M sucrose, but a sucrose concentration of 0.48 M produced a turbidity increase that levelled off after 2 h. Samples from this culture were taken at 30 min intervals and photographed under the phase microscope to demonstrate the sequence of morphological events shown in Fig. 6.

Penicillin-induced lysis of *E. coli* occurred only in a nutritional environment that was capable of supporting bacterial growth. Suspensions of bacteria in media devoid of sources of carbon or nitrogen did not undergo lysis in the presence of penicillin.

Prestidge and Pardee (1957) refined this observation by showing that chloramphenicol, which is a specific inhibitor of protein biosynthesis, protected *E. coli* from the action of 150 μg/ml peni-

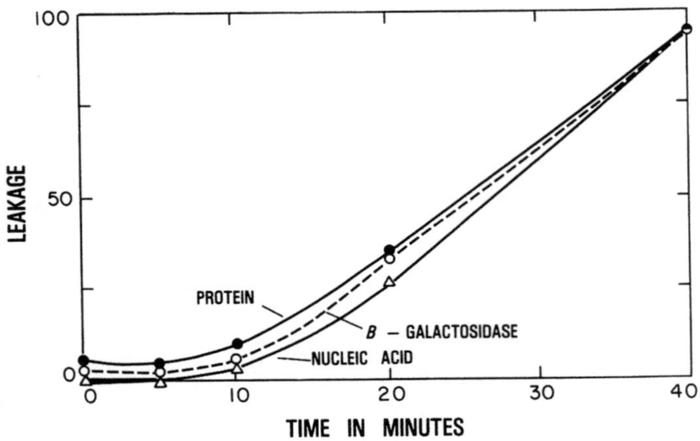

Fig. 9. Release of protein, β-galactosidase and nucleic acids from penicillin-exposed *Escherichia coli* (Prestidge and Pardee 1957)

cillin. When chloramphenicol at 20 μg/ml was supplied at various times to penicillin-containing cultures, the number of bacteria saved from the bactericidal action of penicillin decreased systematically from 52 per cent with chloramphenicol, added at zero time to 12 per cent with chloramphenicol added 20 minutes after penicillin. This work was based on earlier studies of Jawetz et al. (1951) on the interference of chloramphenicol with the action of penicillin.

Prestidge and Pardee did not only present time curves showing the loss of RNA and protein from penicillin-exposed *E. coli*, but also demonstrated the leakage from the bacteria of protein, β-galactosidase and nucleic acids into the experimental medium as a function of time (Fig. 9). They briefly discussed the possbility of a direct action of penicillin on the bacterial membrane, but considered this unlikely because of the observations of Lederberg and Hahn and Ciak which showed that osmotically protected bacteria are converted by penicillin into protoplasts with morphologically intact membranes which in Lederbergs work, after removal of penicillin could partly revert to normal bacillary forms.

H. Interference of Penicillin with Nucleic Acid and Protein Synthesis?

Krampitz and Werkman (1947) and Gros and Macheboeuf (1948) showed that in washed suspensions of certain bacteria, inhibition of RNA dissimilation was caused by concentrations of penicillin 1000 or 10,000 times greater than growth-inhibitory concentrations. Following this, Mitchell and Moyle (1951) proceeded to demonstrate certain imbalances in nuclear acid and free nucleotide and nucleoside composition in *Micrococcus pyogenes* exposed to 50 μg/ml of penicillin in growth medium. These imbalances will not be described in detail. The conclusion is warranted that penicillin does not exert a direct and primary effect on the metabolism of nucleic acids.

For a number of years, beginning in 1947, the idea was entertained that penicillin interfered indirectly or directly with bacterial protein biosynthesis. Gale and his co-workers (Gale and Taylor 1947) reported that Gram-positive bacteria were able to assimilate glutamic acid from the medium in which they were grown and to concentrate this free amino acid within the bacterial cell. They also reported that when certain strains of *Streptococcus faecalis* and *Staphylococcus aureus* were exposed to penicillin during exponential growth, the ability to concentrate free glutamic acid was lost. Gale therefore suggested that "penicillin interferes with the mechanism whereby certain amino acids are taken into the cell, and that the sensitivity of the cell to penicillin is then determined by the degree to which its growth processes are dependent upon assimilation of preformed amino acids rather than upon their synthesis."

However, in 1949, Hunter and Baker obtained a strain of *Bacillus subtilis* which grew readily in a synthetic medium which contained only ammonium sulfate as a source of nitrogen. In this medium, the organism was just as sensitive to penicillin as it was in organic media such as tryptose phosphate broth. They concluded that penicillin inhibited the growth of this strain of *B. subtilis* by some mechanism other than interference with the assimilation of preformed amino acids by the bacterial cell.

One year later, Hotchkiss (1950) published a paper on the abnormal course of bacterial protein synthesis in the presence of penicillin. Washed normal staphylococcal cells, respiring in solutions containing glucose and various mixtures of amino acids, utilized the amino acids and showed an increase in the cellular protein nitrogen. Exposure to penicillin G permitted utilization of oxygen, phosphate and amino acids at essentially the control rates, but there was no increase in the protein nitrogen of the cells. Instead, penicillin-treated cells produced increasing amounts of extracellular substances containing non-amino nitrogen in quantities approximately equivalent to the amino acid nitrogen utilized. This extracellular fraction was tentatively identified as a tetra- or pentapeptide which was produced instead of cellular protein when penicillin was present. Hotchkiss interpreted his observations as indications that penicillin interfered with the conversion of amino acids into staphylococcal protein, i.e. trichloracetic acid insoluble material, in such a manner that extracellular peptides are formed.

Shortly thereafter, Gale and Folkes (1953) discovered that the incorporation of certain amino acids into the trichloracetic acid-insoluble fraction of *S. aureus* was inhibited to the extent of approximately 60 per cent by growth-inhibitory concentrations of penicillin. The inhibition was unusual in that only certain amino acids were affected and that the levels of inhibition reached plateaus which were different for each amino acid. Much later, it became apparent that the amino acids concerned, viz., glycine, glutamic acid, lysine, and alanine are those which occur in the mucopeptide of the bacterial cell wall. So, in fact, Gale, and probably Hotchkiss, were registering amino acid incorporation into the bacterial cell-wall building blocks, and this was, for a while, mistakenly interpreted as an interference with bacterial protein biosynthesis.

I. Park's Nucleotides

Up to this point, I have reported a considerable volume of observations and experimental studies on the action of penicillin as well as various hypotheses. But this entire material did not lead to the recognition of the molecular mechanism of the target reaction whose inhibition is responsible for the bactericidal effect of penicillin. No biochemical reaction of vital importance to the bacterial cell had been demonstrated to be inhibited by bactericidal concentrations of the antibiotic.

It should be remembered that biochemistry during the second part of the 1940's was much concerned with the phosphorylation of metabolic intermediates which was aided by a chapter of Le Page and Umbreit entitled *Methods for the Analysis of Phosphorylated Intermediates* (Umbreit et al. 1945).

Two years later, Park began his studies on the action of penicillin. The original question asked was whether the increase in size of *Staphylococcus aureus*, growing in the presence of 0.1 μg of penicillin, represented an actual increase in cell mass, and especially, whether the various constituents of the bacteria increased at comparable rates (Park and Johnson 1949). The result of the study was that the acid-soluble organic phosphate content of the cells increased at an accelerated rate under the influence of penicillin. This phenomenon was accompanied by a similar increase in a pyrimidine base, tentatively identified as uracil, an increase of pentose identified by the orcinol method, and an increase in reducing material. Measured by dry weight, of nitrogen, phosphorus, and nucleic acid, the cell substance increased by almost 50 per cent in the presence of penicillin. It was inferred that the labile phosphate compound represented a new form of organically bound labile phosphate related to the hypothetical reaction, inhibited by penicillin. The question remained somewhat open whether the reducing material, the material with an absorption maximum of 262 nm, and the labile phosphate were all constituents of a single compound.

One year later a Federation Abstract (Park 1950) followed. It reported that the unknown material was resolved by counter-current distribution into three components. Each of the separated components contained in equimolar proportions uracil, labile phosphate, stable phosphate, pentose and an unknown sugar. The preparation most soluble in the phenol phase contained 3 moles of alanine per mole of labile phosphate and probably one mole of glutamic acid. The second component contained only alanine, while the third component contained no amino acids. The three compounds became known as Park's nucleotides and it was suggested that this series of molecules represented normal intermediates of the cell metabolism and that the inhibition of a specific but

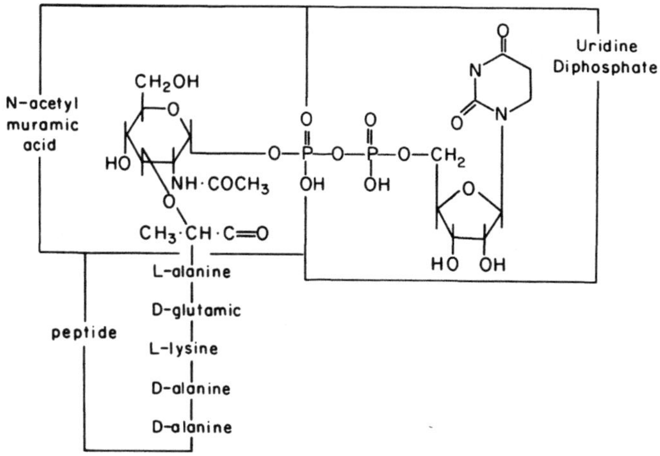

Fig. 10. Park's nucleotide, containing 5 amino acids (Park 1952c)

unidentified reaction by penicillin caused them to accumulate in abnormal amounts.

In 1952 there followed the final series of three companion papers by Park. The first (1952a) presented in detail the isolation of the three compounds from penicillin-exposed *Staphylococcus aureus* and evidence that all three substances were uridine-5-pyrophosphate derivatives. The second paper (1952b) elaborated on the structure common to all three nucleotides, and the third (1952c) contained evidence that the first compound contained no amino acids, the second compound contained one L-alanine residue while the third compound contained a peptide composed of L-lysine, D-glutamic acid and three alanine residues which were a mixture of L-alanine and D-alanine.

Figure 10 shows the definitive structure of the third nucleotide of Park. The structure of the N-acetyl amino sugar was elucidated by Strange (1959), and the final assignment of the stereochemical configurations of the amino acids in the peptide moiety was made by Strominger after the cut-off year of this review.

Until 1957 the role of Park's nucleotides in cell metabolism and in the action of penicillin remained cryptic, although a scholarly lecture by Mitchell (1956) offered a one-sentence speculation that "it is possible that these compounds may be involved in cell envelope synthesis."

The final breakthrough came in 1957, in an article by Park and Strominger, entitled *Mode of action of penicillin: Biochemical basis for the mechanism of action of penicillin and for its selective toxicity*, published in Science (1957). It reviewed the structures of Park's nucleotides and proposed a structure, as shown in Fig. 10, in which only the stereochemical configuration of the three alanine residues remained unspecified. The paper called attention to the fact that the rate of accumulation of the peptide-containing uridine nucleotide in the presence of penicillin suggested that the antibiotic inhibited one of the principal biosynthetic reactions of the bacterial cell.

The paper then reviewed what was known, from the work of others, about the chemical composition of the cell wall and suggested that the uniqueness of the structures in the wall and in the nucleotide means that they must be metabolically related. The reasoning culminated in the conclusion that the accumulation of this compound in penicillin-exposed *Staphylococcus aureus* is a consequence of the interference by penicillin with the biosynthesis of the cell wall. Uridine pyrophosphate glycosyl compounds were regarded as activated intermediates, and the N-acetylamino sugar peptide was regarded as a nucleotidyl fragment, activated for cell wall polymer synthesis. This conclusion was reached 11 years after Duguid had assumed, on the basis of morphological observations, that penicillin was an inhibitor of the formation of the outer supporting cell wall.

Park and Strominger declined to speculate on the exact nature of the interference by penicillin except by referring to the formation of protoplasts as evidence that penicillin interferes with the maintenance of the cell wall or with its synthesis. Duguid had suggested that, except for cell wall biosynthesis, bacterial growth proceeds until the organism finally bursts its defective envelope and so undergoes lysis.

In the same year, 1957, Lederberg in a one-page note expressed some careful doubts in the one-target hypothesis of the action of penicillin. He introduced his comments with a quotation from Eagle and Saz of 1955 "The mechanism whereby penicillin exerts its cytotoxic effect remains obscure." Lederberg emphasized that penicillin-induced bacterial protoplasts, when maintained in an osmotically protective medium, revert to colony-forming bacillary forms when diluted in protective growth media, lacking penicillin. He thought that more remote influences on cell wall formation than those observed by Park and Strominger cannot be precluded and suggested that further studies of antibiotic effects must be conducted with protected protoplasts rather than with lysed or lysing cells in which the ramification of secondary lesions is an inevitable complication."

J. Penicillin Binding by Bacteria

The final section of this review deals with the binding of radioactive penicillin to bacteria and bacterial fractions. This work was started in 1948 by Rowley and his associates and was still in progress when it was briefly and factually reviewed by Eagle and Saz (1955) and discussed in great detail by Cooper (1956).

It is intuitively obvious that in order to affect a living cell, antibiotic molecules must be able to reach and to interact with a vitally important cellular system, the binding site being considered the site of action of the drug. It may represent the molecular machinery operating the biochemical reaction originally inhibited or disorganized by the drug. The study of drug binding to their sites of action is an indispensible part of the investigation of the mode or mechanism of drug action.

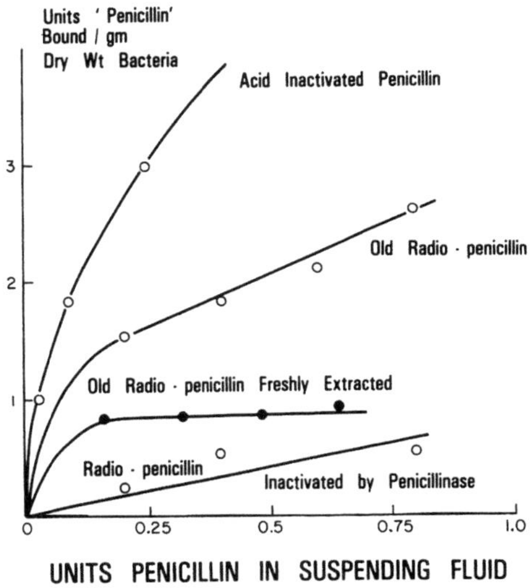

Fig. 11. Bacterial binding of penicillin and degradation products (Cooper et al. 1954)

Penicillin-sensitive strains of bacteria rapidly bound and concentrated the antibiotic with nearly complete equilibration within 1 h at 37°C. With wild strains of bacteria, such as the highly sensitive *Streptococcus pyogenes*, the amount of penicillin bound at concentrations of 0.001-0.1 μg/ml was concentrated up to 200-fold. The binding was specific. It was not observed with penicillamine or penicilloic acid. It also was irreversible. Bound radiopenicillin could not be displaced by fresh non-radioactive penicillin. Extensive washing or treatment with anionic, cationic, or neutral detergents did not liberate bound radiopenicillin. Likewise, pretreatment with such detergents did not affect subsequent binding but pretreatment with non-radioactive penicillin precluded subsequent binding of radiopenicillin. Schepartz and Johnson (1956) reported that alkaline treatment of *Micrococcus pyogenes*, treated with radiopenicillin, resulted in the cleavage of penicilloic acid from the bacteria.

In addition to the specific binding of penicillin, there was non-specific binding of penicillin degradation products. Figure 11 is from a paper of Cooper et al. (1954). The top curve shows the binding by *Staphylococcus aureus* of acid-inactivated penicillin. The second curve, marked old radiopenicillin, demonstrates the progressive binding of ^{35}sulfur from labelled penicillin preparations that had been stored. When this material was repurified by extraction into chloroform at pH 2.3 and back into neutral buffer, the saturation curve through the black dots was obtained. Finally, when this freshly purified radiopenicillin was hydrolyzed with penicillinase, the bottom line was seen. The use of purified radiopenicillin indicated that saturation of the bacterial binding sites was attained at 0.1 μg/ml.

In the same year, 1954, Eagle published an extensive study of radiopenicillin binding to five different bacterial strains. This work used C^{14}- or S^{35}-labelled penicillin. He confirmed that the amount bound from low concentrations was related to the penicillin sensitivity of the strain. Despite wide differences in their sensitivity to penicillin the antibiotic was bound at biologically equivalent levels of 99.9 LDs in comparable amounts. The lethal intracellular concentrations ranged from 1.7 to 4 μg/ml, i.e. 1600 to 3300 molecules penicillin per cell.

Bacteria in the logarithmic phase of growth or suspended in salt solutions, or extracts of bacteria prepared by sonic oscillation had approximately the same reactivity with penicillin. That is, penicillin binding was independent of the metabolic state of the bacteria and was not influenced by differences in the permeability of the cells.

However, at high concentrations of penicillin from 1 to 1000 μg/ml, non-specific additional binding by all the bacterial strains and extracts was observed. This was unrelated to their sensitivity. At these high concentrations, penicilloic acid was bound to the same extent as penicillin itself.

The penicillin binding studies up to the mid 1950's logically led to the asking of two questions, the answers to which remained elusive. The first question concerned the chemical properties of the penicillin-binding component of bacteria. Despite a detailed discussion of experimental studies, Cooper (1956) conceded that "little success has been obtained in characterizing this component, and nothing is known with certainty of its chemical nature." The possibility was considered that the penicillin binding component was located in the osmotic barrier, which is cytologically observed, under the cell wall.

The second question concerned the role of penicillin binding in the antibacterial effects of the antibiotic. It has already been mentioned that Eagle found bacteria to combine with amounts of penicillin far in excess of those which are bound at the bactericidal concentrations. Bacteria suspended in salt solutions combined with penicillin to the same degree as organisms in the logarithmic phase of growth. When such treated bacteria were resuspended in penicillin-free growth medium, they eventually resumed multiplication at normal rates without any release of bound penicillin. It followed that the binding of penicillin alone did not suffice to initiate the bactericidal effect. It is necessary that the cell be in a medium which permits active metabolism and growth and that the antibiotic be continuously present in the surrounding growth medium.

To sum up: during the period on which I have reported, the biochemical theory of the inhibition of cell wall biosynthesis by penicillin emerged as the logical explanation of the antibiotic's mechanism of action but it was not possible to connect this body of knowledge and thought with the experimental results of penicillin binding studies.

References

Abraham, E.P., Fletcher, C.M., Florey, H.W., Gardner, A.D., Heatley, N.G., Jennings, M.A.: Further observations on penicillin. Lancet 177-188 (1941)

Bornstein, S.: Action of penicillin on enterococci and other streptococci. J. Bacteriol. 39, 383-387 (1940)

Bringmann, G.: Elektronenmikroskopische Beobachtungen der Entstehung filtrierbarer (L-) Formen von B. proteus unter Penicillin-Einfluss. Z. Hyg. Infektionskrankh. 135, 557-565 (1952)

Chain, E., Duthie, E.S.: Bactericidal and bacteriolytic action of penicillin in the staphylococcus. Lancet 652-657 (1945)

Chain, E., Gardner, A.D., Heatley, N.G., Jennings, M.A., Orr-Ewing, J., Sanders, A.G.: Penicillin as a chemotherapeutic agent. Lancet 226-228 (1940)

Clutterbuck, P.W., Lovell, R., Raistrick, H.: CCXXVII. Studies in the biochemistry of micro-organisms. XXVI. The formation from glucose by members of the *Penicillium chrysogenum* series of a pigment, an alkali-soluble protein and penicillin — the antibacterial substance of Fleming. Biochem. J. 26, 1907-1918 (1932)

Cooper, P.D.: Site of action of radiopenicillin. Bact. Rev. 20, 28-48 (1956)

Cooper, P.D., Clowes, R.C., Rowley, D.: A note on the use of radioactive penicillin. J. Gen. Microbiol. 10, 246-249 (1954)

Dawson, M.H., Hobby, G.L., Meyer, K., Chaffee, E.: Penicillin as a chemotherapeutic agent. J. Clin. Invest. 20, 434 (1941)

Duguid, J.P.: The sensitivity of bacteria to the action of penicillin. Edinburgh Med. J. 53, 402-412 (1946)

Eagle, H.: Further observations on the zone phenomenon in the bactericidal action of penicillin. J. Bact. 62, 663-668 (1951)

Eagle, H.: The binding of penicillin in relation to its cytotoxic action. I. Correlation between the penicillin sensitivity and combining activity of intact bacteria and cell-free extracts. J. Exp. Med. 99, 207-226 (1954)

Eagle, H., Musselman, A.D.: The rate of bactericidal action of penicillin in vitro as a function of its concentration, and its paradoxically reduced activity at high concentrations against certain organisms. J. Exp. Med. 88, 99-130 (1948)

Eagle, H., Saz, A.K.: Antibiotics. Ann. Rev. Microbiol. 9, 173-226 (1955)

Fleming, A.: On the antibacterial action of cultures of a penicillium with special reference to their use in the isolation of *B. influenzae*. Brit. J. Exp. Pathol. 10, 226-236 (1929)

Flory, H.W., Chain, E., Heatley, N.G., Jennings, M.A., Sanders, A.G., Abraham, E.P., Florey, M.E.: Antibiotics, Vol. II. Oxford: University Press 1949

Gale, E.F., Folkes, J.P.: The assimilation of amino acids by bacteria. 15. Actions of antibiotics on nucleic acid and protein synthesis in *Staphylococcus aureus*. Biochem. J. 53, 493-498 (1953)

Gale, E.F., Taylor, E.S.: The assimilation of amino-acids by bacteria. 5. The action of penicillin in preventing the assimilation of glutamic acid by *Staphylococcus aureus*. J. Gen. Microbiol. 1, 314-326 (1947)

Gardner, A.D.: Morphological effects of penicillin on bacteria. Nature 146, 837-838 (1940)

Gros, F., Macheboeuf, M.: Recherches biochemiques sur le mode d'action de la pénicilline sur un bactérie: *Clostridium sporogenes*. Ann. Inst. Pasteur 74, 368-385 (1948)

Hahn, F.E., Ciak, J.: Penicillin-induced lysis of *Escherichia coli*. Science 125, 119-120 (1957)

Hobby, G.L., Meyer, K., Chaffee, E.: Observations on the mechanism of action of penicillin. Proc. Soc. Exp. Biol. Med. 50, 281-285 (1942)

Hotchkiss, R.D.: The abnormal course of bacterial protein synthesis in the presence of penicillin. J. Exp. Med. 91, 351-364 (1950)

Hunter, T.H., Baker, K.T.: The action of penicillin on *Bacillus subtilis* growing in the absence of amino acids. Science 110, 423-425 (1949)

Jawetz, E., Gunnison, J.B., Speck, R.S., Coleman, V.R.: Studies on antibiotic synergism and antagonism. Arch. Int. Med. 87, 349-359 (1951)

Johnstone, K.I., Crofts, J.E., Evans, D.G.: Single cell culture of Cl. welchii type A morphologically changed by penicillin. Brit. J. Exp. Pathol. 31, 562-565 (1950)

Kochen, M. (ed.): The Growth of Knowledge. New York: Wiley 1967

Krampitz, L.O., Werkman, C.H.: On the mode of action of penicillin. Arch. Biochem. 12, 57-67 (1947)

Lamanna, C., Shapiro, I.M.: Sulfanilamide bacteriostasis in presence of mercuric chloride and p-aminobenzoic acid. J. Bacteriol. 45, 385-394 (1943)

Lederberg, J.: Bacterial protoplasts induced by penicillin. Proc. Natl. Acad. Sci. USA 42, 574-577 (1956)

Lederberg, J.: Mechanism of action of penicillin. J. Bacteriol. 73, 144 (1957)

Lederberg, J., St. Clair, J.: Protoplasts and L-type growth of *Escherichia coli*. J. Bacteriol. 75, 143-160 (1958)

Liebermeister, K., Kellenberger, E.: Studien zur L-Form der Bakterien. I. Die Umwandlung der bazillaren in die globulare Zellform bei Proteus unter Einfluss von Penicillin. Z. Naturforsch. 11b, 200-206 (1956)

Mitchell, P.: Penicillin and the logic of chemotherapy. Giorn. Microbiol. 2, 440-460 (1956)

Mitchell, P., Moyle, J.: Relationships between cell growth, surface properties and nucleic acid production in normal and penicillin-treated *Micrococcus pyogenes*. J. Gen. Microbiol. 5, 421-438 (1951)

Nature 156, 766-767 (1945): Chemistry of Penicillin

Park, J.T.: Isolation of three labile phosphate compounds containing uracil from penicillin-treated *Staphylococcus aureus* cells. Fed. Proc. 9, 213 (1950)

Park, J.T.: Uridine-5'-pyrophosphate derivatives. I. Isolation from *Staphylococcus aureus*. J. Biol. Chem. 194, 877-884 (1952a)

Park, J.T.: Uridine-5'-pyrophosphate derivatives. II. A structure common to three derivatives. J. Biol. Chem. 194, 885-895 (1952b)

Park, J.T.: Uridine-5'-pyrophosphate derivatives. III. Amino acid-containing derivatives. J. Biol. Chem. 194, 897-904 (1952c)

Park, J.T., Johnson, M.J.: Accumulation of labile phosphate in *Staphylococcus aureus* grown in the presence of penicillin. J. Biol. Chem. 179, 585-592 (1949)

Park, J.T., Strominger, J.L.: Mode of action of penicillin. Biochemical basis for the mechanism of action of penicillin and for its selective toxicity. Science 125, 99-101 (1957)

Pratt, R., Dufrenoy, J.: Antibiotics. Philadelphia: Lippincott 1949

Prestidge, L.S., Pardee, A.B.: Induction of bacterial lysis by penicillin. J. Bacteriol. 74, 48-59 (1957)

Reid, R.D.: Some properties of a bacterial-inhibitory substance produced by a mold. J. Bacteriol. 29, 215-221 (1935)

Rowley, D., Miller, J., Rowlands, S., Lester-Smith, E.: Studies with radioactive penicillin. Nature 161, 1009-1010 (1948)

Science 102, 627-629 (1945): Chemistry of Penicillin

Schepartz, S.A., Johnson, M.J.: The nature of the binding of penicillin to bacterial cells. J. Bacteriol. 71, 84-90 (1956)

Stent, G.S.: Prematurity and uniqueness in scientific discovery. Sci. Am. 227 (6), 84 (1972)

Strange, R.E., Kent, L.H.: The isolation, characterization and chemical synthesis of muramic acid. Biochem. J. 71, 333-339 (1959)
Umbreit, W.W., Burris, R.H., Stauffer, J.F.: Manometric techniques and relared methods for the study of tissue metabolism. Minneapolis: Burges 1945
Wilson, D.: In Search of Penicillin, pp. 111-115. New York: Knopf 1976

The Role of Inducible DNA Repair in W-Reactivation and Related Phenomena*

Gregory B. Knudson

A. Introduction

Damage to cellular DNA or inhibition of DNA replication induces an array of coordinately controlled functions ("SOS" responses) in *Escherichia coli* including derepression of repair enzymes and mutagenesis. Most of the UV-induced photoproducts in the DNA of bacteria and their viruses are repaired by "error-proof" repair mechanisms such as photoreactivation, "short-patch" excision repair, and recombinational repair. Single-strand gaps resulting from replication of DNA which contains pyrimidine dimers, can be repaired by ultraviolet (UV)-inducible, *recA*$^+$ *lexA*$^+$-dependent "error-prone" repair ("SOS" repair) which leads to mutations in the bacteria and its phages. Other UV-induced, *recA*$^+$ *lexA*$^+$-dependent functions, such as prophage induction, filamentous growth, and W-reactivation, are coordinately regulated with the error-prone repair functions. An increasing amount of experimental evidence suggests that misincorporation of bases could result from an inducible factor which affects the proofreading activities of DNA polymerase, allowing it to fill postreplication daughter strand gaps across from noncoding UV photoproducts. Mutagenic W-reactivation illustrates the role of UV-inducible, *recA*$^+$ *lexA*$^+$-dependent functions in the error-prone DNA repair of bacteria and their viruses.

B. Enzymatic Repair Systems in Bacteria and Their Viruses

Ultraviolet irradiation of bacteria damages the DNA by forming intrastrand pyrimidine dimers (Setlow and Carrier 1966) and other lesions which distort the DNA duplex (Marmur and Grossman 1961). These events may ultimately lead to cell death or to mutations through error-prone repair. In UV-irradiated *E. coli* about six pyrimidine dimers are formed per genome per erg per mm^2 (Witkin 1969c). Bacteria can remove these UV-induced distortions in the DNA double helix by several enzyme repair systems, including photoreactivation, "short-patch" and "long-patch" excision repair (dark repair), pre- and postreplication recombinational repair, and inducible error-prone ("SOS") repair. Some bacteriophages possess genes for enzymatic dark re-

---

*The views of the author do not purport to reflect the positions of the Department of the Army or the Department of Defense

Progress in Molecular and Subcellular Biology, Vol. 8, edited by F.E. Hahn

pair operating to minimize the damaging effects of UV-induced lesions in their DNA (Hayes 1974) thereby protecting the integrity of their genetic material. Ultraviolet-induced lesions in the DNA of infecting phages that do not have their own DNA repair systems may be repaired by the bacterial host systems. Thus the lethal effects of dimers in the phage DNA can be overcome by (a) their monomerization through enzymatic photoreactivation (Dulbecco 1950); (b) the physical removal by an excision-repair system, as in "host cell reactivation;" (c) genetic recombination, as in "prophage reactivation" and "multiplicity reactivation" (Luria 1952; Bernstein 1981); or (d) the inducible error-prone repair system, as in "Weigle-reactivation" (UV-reactivation).

The most important biological effect of UV irradiation is the photochemical formation of thymine dimers in the DNA double helix. Kelner (1949) first observed that the killing and mutagenic effects of ultraviolet light can be prevented by exposing UV-irradiated *E. coli* to visible light. Wulff and Rupert (1962) demonstrated that photoreactivation results from the error-free monomerization of pyrimidine dimers in situ by a photoreactivating enzyme which is activated by visible light. The ability to hydrolyze thymine dimers photochemically can be lost by a single *phr*$^-$ gene mutation (Setlow and Setlow 1963; Hanawalt 1968). This enzyme, isolated from the photoreactivable species *E. coli* or *Saccharomyces cerevisiae*, can restore about 10% of the transforming activity to UV-inactivated transforming DNA in the presence of visible light of wavelength 300-400 nm (Rupert 1961). *Bacillus subtilis*, *Pneumococcus*, and *Haemophilus* lack this enzyme and are not photoreactivable (Hayes 1974).

Garen and Zinder (1955) found that some UV-irradiated bacteriophages could be reactivated by nonirradiated host bacteria. This mode of UV repair, later named host cell reactivation (HCR), was shown to result from excision of the regions containing pyrimidine dimers in the phage followed by repair synthesis (Boyce and Howard-Flanders 1964; Setlow and Carrier 1964; Devoret et al. 1975). In host cell reactivation, three widely separated genes on the chromosome of *E. coli* K12, *uvrA*$^+$, *uvrB*$^+$, and *uvrC*$^+$ (Van de Putte et al. 1965; Mattern et al. 1965), control the repair of UV lesions in the infecting phage genome or in the bacterial DNA resulting from the excision of pyrimidine dimers. A mutation in any one of these three unlinked loci (*uvrA*, *uvrB*, or *uvrC*) shows a ten- to twenty-fold increase in sensitivity to UV (Lewin 1974) and cannot repair UV damage to infecting phages (Hcr$^-$) or their own genomes due to their inability to excise pyrimidine dimers from DNA (Howard-Flanders et al. 1966). The *uvr*$^-$ strains retain most of their resistance to X-rays and other ionizing radiation and show normal recombination (Hayes 1974). Excision repair of DNA exposed to UV has been described at the biochemical level as a cut-and-patch process. DNA polymerase I and the *uvrA*, *uvrB* and *uvrC* proteins bind to and nick the 5' end of the damaged site (Grossman et al. 1975). Single stranded oligonucleotides which include the dimer and a small number of bases on either side of it are excised followed by gap enlargement and repair synthesis utilizing the bases opposite the excised segment as a template. The sugar-phosphate backbone is then sealed by a DNA ligase (Kushner et al. 1971). This "short-patch" excision repair eliminates photodimers by bacterial Uvr^+ functions before DNA replication and does not seem to be mutagenic (Defais et al. 1971). "Long-patch" excision repair, which requires the *recA*$^+$ *lexA*$^+$ genotype, introduces errors in the repair of excision gaps at a low frequency.

The *uvrA* and *uvrB* excision repair enzymes are synthesized at increased levels following UV-induced damage to the DNA or inhibition of DNA replication. These *uvrA* and *uvrB* genes are part of an inducible regulon controlled by the *recA* and *lexA* gene products. Both *uvrA* and *uvrB* genes are repressed by *lexA* protein binding to their operator regions and are induced by the proteolytic inactivation of the *lexA* protein in response to DNA damage (Fogliano and Schendel 1981; Kenyon and Walker 1981).

C. W-Reactivation and W-Mutagenesis

UV-reactivation is one of several interacting reactions to primary DNA damage. Weigle (1953) experimentally defined "UV-restoration" as the increased survival of UV-irradiated phage λ when plated on *E. coli* host cells stimulated by slight UV-irradiation before phage infection or by UV-irradiation of the phage-bacterium complexes after adsorption of the UV-inactivated phage. "UV-reactivation" (UVR) has come to refer to the higher survival of UV-irradiated bacteriophage in general when it infects host cells that have been exposed to a low dose of UV irradiation before infection, as compared with infection of nonirradiated host cells. Among the reactivated phages, a fairly large proportion are mutants. This high rate of mutation among the reactivated phage is referred to as "UV-mutagenesis." Since ultraviolet light is not the only agent that can induce this kind of reaction, the terms "W-reactivation" (Weigle-reactivation) and "W-mutagenesis" (Weigle-mutagenesis) are used in place of the older terms "UV-reactivation" and "UV-mutagenesis" to describe the more general phenomena of error-prone repair of phage lesions that occur when host bacteria are stimulated by exposure to a mutagen prior to phage infection (Radman 1974).

W-reactivation was first observed as the increased survival of UV-irradiated phage λ when infecting UV-irradiated *E. coli*, as compared with infection of nonirradiated *E. coli* (Weigle 1953). The W-reactivation process has been demonstrated in several other *E. coli* bacteriophages including T3 (Weigle and Dulbecco 1953); P22 (Garen and Zinder 1955); T1 (Garen and Zinder 1955; Tessman 1956); P1 (Kerr and Hart 1973); P2 (Bertani 1960); HP1 (Harm and Rupert 1963); and T7 (McKee and Hart 1975); and in several bacteriophage containing single-stranded DNA including S13 (Tessmand and Ozaki 1960); φR (Ono and Shimazu 1966); and φX174 (Das Gupta and Poddar 1975). W-reactivation was also demonstrated with other UV-irradiated phages including AR-1 and SP-50 plated on UV-irradiated *B. subtilis* (Azizbekyan and Galitskaya 1975); phage 81A in *Streptococcus faecalis* X14 (Miehl et al. 1980); phage P22 in *Salmonella typhimurium* TA92

(Walker 1978); and phages PS20 and 368ϕ in *Serratia marcescens* (pKM101; R68.45) (Knudson 1977).

W-reactivation of UV-irradiated phage λ can be initiated not only by exposure of the *E. coli* host cells to UV, but also by their exposure to nitrogen mustard (Weigle 1953); X-rays (Weigle 1953; Ono and Shimazu 1966); mitomycin C (Otsuji and Okubo 1960); or to nonlethal periods of thymine starvation (Hart and Ellison 1970). W-reactivation of UV-irradiated phage λ also occurs in *E. coli* with DNA ligase deficiency (Morse and Pauling 1975) or with the presence of the *tif*$^-$ mutation following a thermal shift (Castellazzi et al. 1972a). Phage damaged by agents other than UV-irradiation also show W-reactivation, such as gamma radiation (Bresler et al. 1978), nitrous acid (Ono and Shimazu 1966), hydroxylamine (Vizdalova 1969), nitrogen mustard (Kerr and Hart 1972), and 5-bromouracil incorporation before UV-irradiation (Kneser et al. 1965; Kerr and Hart 1972).

The variety of DNA lesions which can be W-reactivated indicates that the process must be of a generalized nature, depending on elimination of the damage rather than its reversal. Treatments that promote W-reactivation of phage λ in nonlysogenic *E. coli*, such as UV-irradiation, thymineless death, or thermal shift of *tif*$^-$ mutants, also promote lysogenic induction in lysogenic bacteria (Hart and Ellison 1970; Castellazzi et al. 1972a).

Several possible mechanisms for W-reactivation have been postulated. Harm (1963) reported that W-reactivation is not a specific repair process but an enhancement of host cell reactivation. He showed that in the "completely" HCR-deficient strain of *E. coli* K12s (*hcr*$^-$), which inhibits the excision of pyrimidine dimers from both bacterial and phage DNA mediated by host-cell enzymes, no W-reactivation is found. W-reactivation took place in systems which showed HCR and was absent in those where no HCR was observed. In support of this hypothesis, Mattern et al. (1965) and Ogawa et al. (1968) reported that the UV-sensitive Hcr$^-$ strains (mutants in *uvrA*, *uvrB*, or *uvrC*) of *E. coli* lack the capacity for both host cell reactivation and W-reactivation. However, in contradiction to Harm's (1963) hypothesis, Kneser et al. (1965) and Kneser (1968) reported that even when HCR was efficiently restricted by using the nonhost cell-reactivating strain of *E. coli* K12s (*hcr*$^-$) as the host, appreciable W-reactivation of λ phage was observed. Harm (1966) defended his hypothesis that increased phage survival under W-reactivation conditions is due to enhanced efficiency of HCR by showing that Hcr$^-$ bacteria still exhibit "residual" HCR. So the W-reactivation of phage λ observed in *E. coli* K12s (hcr$^-$) was said to be due to "residual" HCR but not "ordinary" HCR (Kneser 1968). This "residual" HCR which allows W-reactivation in an *E. coli hcr*$^-$ strain may be coded for by the gene *uvrD* (Ogawa et al. 1968), while *hcr*$^-$ mutants at the *uvrA*, *uvrB*, and *uvrC* loci block "ordinary" HCR (Rupert and Harm 1966). Since three known types of excision-deficient mutants of *E. coli* K12(*uvrA*$^-$, *uvrB*$^-$, and *uvrC*$^-$) exhibit W-reactivation of λ phage with approximately the same efficiency of W-reactivation in the excision-proficient and -deficient strains, Radman and Devoret (1971) and Defais et al. (1971) suggested that W-reactivation does not involve excision repair of pyrimidine dimers in phage DNA. Boyle and Setlow (1970) also reported that W-reactivation of λ phage does not involve HCR and that it is achieved by different mechanisms of repair. W-reactivation of λ phage in *uvr*⁻ mutants strongly depends on the UV dose given to both the phage and to the host cells (Defais et al. 1971), which is probably why W-reactivation in excision-deficient mutants was not detected in earlier investigations. Defais et al. (1971) further suggested that the *uvr*⁺ functions of excision repair, which is an accurate repair mechanism (Witkin 1969a,b), are not required for W-mutagenesis of λ phage. The single-stranded DNA phage ϕR fails to be reactivated by the HCR mechanism of *E. coli* but is competent for W-reactivation (Ono and Shimazu 1966), which further suggests that W-reactivation and host cell reactivation are due to independent mechanisms.

In UV-irradiated *E. coli*, both excision and resynthesis are carried out by DNA polymerase I (Hanawalt and Setlow 1975). Ogawa (1970) reported that the *polA*⁻$_1$ mutant of *E. coli* K12, which is deficient in DNA polymerase activity and has increased UV-sensitivity, lacks W-reactivation ability for UV-irradiated λ phage. Paterson et al. (1971) also reported that *polA*⁻$_1$ mutants are almost totally deficient in W-reactivation. However, Caillet-Fauquet and Defais (1972) showed that maximum W-reactivation occurs in *polA*⁻ mutants, as in *uvrA*⁻ mutants, at lower UV doses than in the wild type cells. Ogawa (1970) and Paterson et al. (1971) did not observe W-reactivation in the *polA*⁻ mutants due to the high doses of UV used on both λ and the host cells.

Defais et al. (1971) suggested that the UV-induced repair system resulting in W-reactivation and W-mutagenesis acts directly on unexcised dimers in λ DNA. This was consistent with the finding of Tomilin and Mosevitskaya (1975) which showed that the UV-endonuclease from *Micrococcus luteus* strongly decreases W-reactivation and mutagenic (error-prone) repair of UV-irradiated λ DNA. Roulland-Dussoix (1967) and Boyle and Setlow (1970), suggested that W-reactivation results from the protection of the UV-damaged phage DNA against attack by nucleases which are directed against UV-damaged bacterial DNA. W-reactivation could occur because UV-damaged host DNA acts as a substrate for the excision enzymes competing for their action with dimers in the λ phage DNA and shifting the balance in favor of phage repair. Harm and Rupert (1963) also interpreted W-reactivation to be the result of competition between irradiated bacterial DNA and phage DNA for a nuclease specific for UV damage. However, this hypothesis does not explain how the remaining pyrimidine dimers in the λ phage DNA are eliminated or bypassed. Kneser et al. (1965) criticized the nuclease-protection theory on the grounds that W-reactivation of phage can occur when HCR is inhibited in *uvr*⁻ strains.

Mutants of *E. coli* K12 which have a Rec⁻ phenotype are defective in genetic recombination, highly sensitive to UV and X-irradiation and map at three separate loci (Clark and Margulies 1965; Barbour and Clark 1970). The radiation-sensitive, recombination deficient *recA*⁻ mutants degrade an abnormally large amount of their genome when exposed to UV light, earning them the name "reckless" mutants (Howard-Flanders and Theriot 1966). The closely linked, radiation-sensitive, recombination deficient *recB*⁻ and *recC*⁻ "cautious" mutants degrade their DNA much less after UV-irradiation (Lewin 1974). The *recB*⁺ and *recC*⁺ gene products may be nucleases which are responsible for much of the DNA breakdown following UV-irradiation, while the *recA*⁺ gene product may limit the process in some way.

Bacterial *recA*⁺ function is necessary for W-reactivation of UV-damaged phage λ (Ogawa et al. 1968; Shimada et al. 1968). The *recA*⁺ gene function is also required for W-mutagenesis (the increased frequency of mutations in the surviving phage from W-reactivation) and is eliminated by a single *recA*⁻ mutation (Miura and Tomizawa 1968). Both W-reactivation and W-mutagenesis of phage λ are independent of *recB*⁺ and *recC*⁺ gene products (Kerr and Hart 1972). McKee and Hart (1975) showed that the *recA*⁺ function is needed for W-reactivation of T7; Das Gupta and Poddar (1975) reported that the *recA*⁺ gene is essential for W-reactivation of the single stranded DNA phage ϕX174. W-reactivation and W-mutagenesis of phage λ also require the bacterial *lexA*⁺ gene function (Defais et al. 1971) in *E. coli* K12 and the analogous *exr*⁺ gene function in *E. coli* B (Witkin 1967a). W-reactivation does occur in the absence of *recA*⁺ and *exr*⁺ alleles when phage λ is damaged by nitrous acid, hydroxylamine, or nitrogen mustard, rather than by UV (Kerr and Hart 1972). The *recA*⁺ and *exr*⁺ gene products may be necessary to stabilize or modify UV lesions in phage DNA, but not other kinds of lesions before W-reactivation can take place, rather than being directly involved in the basic repair process itself. UV-induction of the prophage λ (Donch et al. 1970; Castellazzi et al. 1972b; Mount et al. 1976) also require the *recA*⁺ and *lexA*⁺ functions. Host cell reactivation is also less effective in *recA*⁻ mutants (Echols and Gingery 1968), suggesting that some common pathways are involved in these phenomena.

As replication proceeds in UV-damaged bacterial DNA, unrepaired pyrimidine dimers cause the formation of daughter strand gaps in the newly formed DNA strand opposite each unrepaired lesion (Howard-Flanders et al. 1966, 1968; Rupp and Howard-Flanders 1968). These discontinuities in the newly replicated molecule, which are lethal if not repaired, cannot be filled by excission repair since they are located opposite a non-coding lesion. Postreplication repair of gaps opposite unexcised pyrimidine dimers can proceed by a *recA*⁺-dependent recombinational process between complementary daughter strands in which the intact region of one daughter strand serves as a template for repair of the gap in the other to restore the original DNA structure (Rupp et al. 1971). According to Witkin (1969c), UV-induced mutagenesis results from errors due to inaccurate postreplication repair dependent upon the *lexA*⁺ function.

The *recA* gene, which has recently been sequenced, encodes a single protein with a molecular weight of 37,842 (Horii et al. 1980; Sancar et al. 1980). *E. coli recA* protein binds to single-strand gaps and catalyzes ATP-dependent hybridization of the single-strand regions to homologous regions in the intact sister chromatid (McEntee et al. 1979; Shibata et al. 1979). The *recA* protein then promotes strand-exchange activity, driven by ATP hydrolysis (Cassuto et al. 1980; West et al. 1981). In this way, the postreplication gaps in the daughter strand can be filled by recombination with the double-stranded sister duplex. When complexed with single-strand regions of the DNA, the *recA* protein also functions as a protease and cleaves the *lexA* protein, which is a repressor of the *recA* gene (Gudas and Mount 1977) and other unlinked target genes in the "SOS" system. The *recA* protein which is normally produced at a low basal level, increases to several thousand *recA* protein molecules per cell following DNA damage.

In prereplication recombinational repair, which involves recombination of two DNA duplexes, there is no need for DNA synthesis since all the DNA strands needed for recombination are already present. Prereplication recombinational repair is a nonmutagenic (error-free) process.

In prophage reactivation, first described by Jacob and Wollman (1953), the survival of UV-irradiated phage λ is greater when plated on a bacterium carrying a homologous prophage than on a host nonlysogenic or lysogenic with a nonhomologous prophage. Prophage reactivation also occurs in phage P2 and phage P22 (Yamamoto 1967). This type of repair is due to prereplication recombination between the homologous DNA of the UV-damaged phage and the intact resident prophage (Yamamoto 1967), so there is no need for any extensive DNA synthesis. This process is not mutagenic and is strongly decreased in a host carrying $recA^-$ mutations, but is not influenced by the presence or absence of the $lexA^+$ function (Blanco and Devoret 1973).

Garen and Zinder (1955) proposed the hypothesis that in W-reactivation, UV-irradiation of *E. coli* stimulates recombinational repair between the homologous regions of UV-damaged phage DNA and intact bacterial DNA, resulting in increased viability of the irradiated phage. This hypothesis could explain the requirement of recombination functions for UV-reactivation (Weigle 1966). Hart and Ellison (1970) proposed that W-reactivation and prohage reactivation of phage λ occur by related mechanisms since UV- and nitrous acid-damaged phage are reactivated by both processes. Furthermore, both processes are eliminated by $recA^-$ mutants in the host bacterium while neither process is appreciably affected by *hcr* alleles. W-mutagenesis could be the result of inaccurate recombination or imperfect homology between the phage and bacterial DNA (Tessman 1956; Stent 1958).

Other experimental results do not support the idea that W-reactivation occurs as a result of prereplication recombination between homologous parts of the UV-irradiated phage and bacterial chromosomes (Ogawa and Tomizawa 1973). Radman and Devoret (1971) reported that the absence of attachment regions in the host and λ chromosomes does not effect W-reactivation. Blanco and Devoret (1973) showed that W-reactivation of phage λ does not require, and is not enhanced by, the presence of large pieces of DNA which are homologous to the infecting phage DNA. The independence of W-reactivation and prereplication recombination is further demonstrated by the fact that UV-reactivation requires the $lexA^+$ function (Defais et al. 1971) and is highly mutagenic, while prophage reactivation is unaffected by mutations in *lexA* and is not mutagenic (Blanco and Devoret 1973). In contrast to prereplication recombinational repair, W-reactivation is unaffected by the *recB* and *recC* genes and the need for the $recA^+$ and $lexA^+$ gene products in W-reactivation is dependent on the nature of the lesion in the DNA (Kerr and Hart 1972). W-reactivation occurs in the single stranded DNA phage S13, ϕR, and ϕX174, which seems to exclude the possibility that W-reactivation is due to a prereplication recombinational repair mechanism between two DNA duplexes, as in prophage reactivation. The "recombination" hypothesis as the mechanism of W-reactivation presumes a high degree of homology between the bacterial and the phage DNA. However, in the DNA of phage AR9, which is subject to W-reactivation in *B. subtilis*, the thymine has been replaced by hydroxymethyluracil. Furthermore, the compositions of the DNA bases of the phage and bacteria differ substantially [G + C = 27.7 and 44.8, respectively (Azizbekyan and Galitskaya 1975)]. These results indicate that prophage reactivation and W-reactivation are engendered by two different mechanisms.

Although UV-reactivation and UV-mutability of λ phage are correlated, in that they are both blocked by $recA^-$ and $lexA^-$ (uninducible repressor) mutations, are both UV-induced, and show comparable kinetics as a function of UV dose given to the host cells and to the phage (Defais et al. 1971; Wackernagel and Winkler 1971; Mount et al. 1972), there is evidence that they are separate, non-interdependent phenomena. Kerr and Hart (1972) showed that when phage λ is damaged by nitrous acid, hydroxylamine, or nitrogen mustard, rather than by UV, W-reactivation does occur in $recA^-$ and exr^- mutants. However, W-mutagenesis of λ phage, which normally accompanies W-reactivation, does not occur in the $recA^-$ or exr^- bacteria (Kerr and Hart 1972). This demonstrates that W-mutagenesis of λ phage is not an essential feature of the W-reactivation process itself. Although W-reactivation is independent of the $recA^+$ and exr^+ alleles, W-mutagenesis is not, and like UV-induced error-prone repair mutagenesis of bacteria, it does not occur in $recA^-$ or exr^- mutants. Morse and Pauling (1975) have shown that W-mutagenesis in *E. coli* is not an essential component of W-reactivation because phage mutagenesis, but not reactivation, is inhibited by 3 μg/ml of chloramphenicol. Mount et al. (1976) reported that phage λ is W-reactivated in the tsl^- $recA^-$ mutant strain of *E. coli* with about one-half the efficiency of that in the wild type strain, but with no corresponding mutagenesis of the phage. Rothman et al. (1979) showed that *E. coli recF*143 *uvrB*5 mutant strains completely block all W-reactivation but permit normal levels of mutagenesis. This suggests that there is a UV-inducible error-free pathway of DNA repair in *E. coli* that is independent of the UV-inducible error-prone repair of phage lesions.

George et al. (1974) demonstrated indirect W-reactivation of phage λ and its mutagenesis by introducing UV-damaged DNA into a nonirradiated F^- recipient host cell through conjugation with a UV-irradiated Hfr (high-frequency recombination) or F-lac^+ donor prior to infection with UV-irradiated phage. George et al. (1974) found that chromosomal or episomal UV-damaged DNA is effective in inducing indirect W-reactivation even in uvr^- strains, in which there is no excision of pyrimidine dimers.

D. Direct and Indirect Prophage Induction

The lysogenic state of prohage λ is maintained by a cytoplasmic protein repressor which is specified by the c_1 region of the λ phage genome (Ptashne 1967). Lwoff et al. (1950) first demonstrated that treating bacteria with small doses of UV can induce the prophage to enter the vegetative state, multiply, and lyse the bacteria (direct induction). Jacob and Wollman (1959) showed that X-rays and nitrogen mustard were also effective inducing agents. Treatment of *E. coli* K12 lysogenic for phage λ with mitomycin C or deprivation of thymine, which inhibits bacterial DNA synthesis, also induces prophage λ to enter the vegetative state (Korn and Weissbach 1962).

Indirect induction, first described by Borek and Ryan (1958), occurs when an $F^+\lambda^-$ donor *E. coli* K12 is irradiated with UV and crossed with a non-irradiated F^- cell lysogenic for λ. The indirect induction of prohage λ is mediated through the UV-irradiated sex factor. However, unlike indirect W-reactivation, indirect induction does not occur when UV damaged chromosomal DNA from an Hfr λ^- cell is introduced by conjugation into an $F^-\lambda^+$ recipient (Devoret and George 1967). Indirect induction can also result from the conjugal transfer of UV damaged F', such as F-lac$^+$ and F-gal$^+$, into F^- lysogenic recipients (George and Devoret 1971). Monk (1969) reported that indirect induction of prophage λ in a lysogenic recipient cell can be brought about by conjugation with a UV-irradiated donor cell carrying the transmissible Col I factor (colicinogen) or RTF (resistant transfer factors). Infection with irradiated phage P1, which is maintained in an autonomous state (Ikeda and Tomizawa 1968) also brings about indirect induction of λ lysogens (Rosner et al. 1968). The replicons Col I, RTF, the sex factors F and F', and the phage P1, all of which can mediate indirect induction of a lysogen, have in common the ability to replicate coordinately with the bacterial chromosome without integrating into the bacterial chromosome.

Recombination-deficient mutants (*recA*$^-$) of *E. coli* K12 which are lysogenic for λ phage are not inducible by UV irradiation (Fuerst and Simonovitch 1965; Hertman and Luria 1967). Indirect induction, mediated by irradiated F or P1, does not occur in *recA*$^-$ mutants which are lysogenic for λ (Brooks and Clark 1967; Rosner et al. 1968). When the *rec*$^+$ gene is introduced into *E. coli* K12 *rec*$^-$ (λ^+) by transduction with phage P1, it is then capable of producing λ phage after UV induction (Hertman and Luria 1967). This shows that the failure of UV to induce λ in the *rec*$^-$ lysogens is due to an inability to lift repression rather than to damaged prophage. Direct UV-induction of λ, like W-reactivation, also requires the *lexA*$^+$ gene in *E. coli* K12 (Castellazzi et al. 1972b) or the equivalent *exr*$^+$ gene in *E. coli* B (Donch et al. 1970).

In an ATP-dependent reaction, the *recA* protein becomes an activated protease by complexing with single-strand DNA opposite gaps which are generated by replication past unexcised dimers.

The activated *recA* protein-DNA complex inactivates the phage λ repressor, and other repressors, by proteolytic cleavage. In this way, damage to DNA sufficient to generate postreplication single-strand gaps activates the *recA* protein which cleaves and inactivates the λ repressor resulting in lysogenic induction (Roberts et al. 1978; Craig and Roberts 1980). Although the *lexA* repressor is the primary substrate of the activated *recA* protease, there is a selective advantage for phage λ to have a repressor that is also sensitive to this protease. In this way the prophage can escape from a cell whose DNA has been severely damaged and that may die (Little et al. 1980).

E. Plasmid Mediated UV-Protection and Mutagenesis

Plasmid mediated resistance to killing by UV-irradiation (UV-protection) with an increase in mutagenesis has been reported in *S. typhimurium* and *E. coli* containing Col I (Howarth 1965, 1966), R46 (Mortelmans and Stocker 1976) and its derivative pKM101 (Walker 1977; Walker and Dobson 1979) and R-Utrecht (MacPhee 1973); in *E. coli* (Monti-Bragadin et al. 1978; Siccardi 1969) and *Pseudomonas aeruginosa* (Krishnapillai 1975; Lehrbach et al. 1977, 1978) containing R factors and sex factors; in *S. faecalis* (Miehl et al. 1980); and in *S. marcescens* containing plasmids pKM101 and R68.45 (Knudson 1977). Plasmid R46 mediated UV-protection and enhanced UV-mutagenesis are absolutely dependent on the $recA^+$ genotype in *E. coli* and *S. typhimurium* (Mortelmans and Stocker 1976) and dependent on the host $lexA^+$ function in some strains of *E. coli* (Waleh and Stocker 1979). The ability of plasmid pKM101 to enhance mutagenesis and DNA repair is $recA^+$ $lexA^+$-dependent in *S. typhimurium* (McCann et al. 1975). Plasmid R46 and pKM101 mediated W-reactivation of UV-irradiated phage λ in *E. coli* (Walker 1977) and phage P22 in *S. typhimurium* (Walker 1978) has also been observed. The evidence suggests that these plasmids amplify the activity of the inducible error-prone repair system of the host.

F. Inducible "SOS"-Repair Functions

W-reactivation, W-mutagenesis, direct and indirect induction of prophage λ, cell filamentation and UV-mutagenesis of bacteria are all related phenomena with common pathways (Sedgwick 1975a, b). None of these phenomena are expressed constitutively but all are induced by UV. W-reactivation and W-mutagenesis of irradiated bacteriophage, induction of phage λ, and UV-mutagenesis of bacteria resulting from error-prone repair are dependent on $recA^+$ and $lexA^+$ (equivalent to exr^+) gene functions (Miura and Tomizawa 1968; Mount et al. 1976). Filamentous growth in *lon* strains, which is strikingly parallel to prophage induction (Witkin 1967b), has the same requirement for $recA^+$ (Green et al. 1969) and $lexA^+$ functions (Donch et al. 1968). Long filaments can also be induced indirectly by mating UV-irradiated Col I donor cells with nonlysogenic (λ^-) recipients (Kirby et al. 1967; Monk 1969). All of these UV-inducible, *recA*$^+$ *lexA*$^+$-requiring functions are expressed constitutively in *E. coli* K12 *tif* mutants following a thermal shift to 42°C for 45 min (Castellazzi et al. 1972a; Witkin 1974). The *tif* gene, which is an allele of *recA*, may code for an altered *recA* protein that is activated spontaneously, at the elevated temperature, in the absence of single-strand DNA cofactor (Mount 1977). These effects are abolished if chloramphenicol is present during incubation at the elevated temperature, just as in the case of prophage induction (Witkin 1975).

Protein and RNA synthesis is essential for W-reactivation (Ono and Shimazu 1966). Protein synthesis is also necessary for UV-mutagenesis (Witkin 1956) and for UV-induction of prophage λ (Tomizawa and Ogawa 1967). UV-inducible, *recA*$^+$ *lexA*$^+$-requiring functions can be coordinately induced in response to the inhibition of DNA synthesis, while RNA and protein synthesis continues (Witkin and George 1973; Witkin 1974, 1975). Monk and Kinross (1975) showed that the arrest of DNA synthesis is essential for λ induction. *E. coli lig*$^-$ mutants produce defective DNA ligase, which is an enzyme that closes single-strand nicks in the DNA duplex resulting from UV-irradiation (Bonura and Smith 1975). DNA ligase deficiency which results in the inhibition of DNA synthesis (Pauling and Hamm 1969) leads to the stimulation of prophage induction (Gottesman et al. 1973), W-reactivation, and W-mutagenesis by depressing the error-prone repair pathway (Morse and Pauling 1975).

Radman (1974) has called this *recA*$^+$ *lexA*$^+$-dependent error-prone repair pathway, which is inducible by DNA lesions which temporarily block DNA replication, "SOS" repair. We know that in *E. coli* this repair pathway is (a) UV-inducible, (b) requires the *recA*$^+$ and *lexA*$^+$ functions, (c) requires protein synthesis without DNA synthesis, (d) is error-prone, (e) involves a repair mechanism different from pre- or postreplication recombinational repair or excision repair processes, and (f) is common to the phenomena of W-reactivation, W-mutagenesis, UV-induction of λ, cell filamentation, and UV-mutagenesis of bacteria (Devoret 1973). Witkin (1974, 1976) suggested that all of the UV-inducible *recA*$^+$ *lexA*$^+$-dependent functions express the activity of newly derepressed genes, governed by repressors which respond to a common effector produced by a common induction pathway. The inactivation of a repressor has been demonstrated in prophage induction. Inhibition of DNA synthesis by UV lesions may initiate the induction pathway. The UV-inducible product may inhibit the 3'-5'-exonuclease "proofreading" activity of DNA polymerase. A DNA polymerase with relaxed template dependence could polymerize DNA past noninstructive UV photoproducts, while filling postreplication daughter strand gaps, with a high probability of inserting "wrong" bases.

Inhibition of DNA replication or damage to the DNA which is sufficient to produce postreplication gaps initiates the coordinated expression of the pleiotropic "SOS" response which aids cell survival by enhancing DNA repair capacity. The *recA* protein bids to single-strand DNA opposite gaps; this binding activates its proteolytic activity. The activated *recA* protein mediates the inactivation and cleavage of the repressor encoded by the *lexA* gene (Little et al. 1980) and the cleavage of other repressors, including phage λ and phage P22 repressors (Phizicky and Roberts 1980). The *lexA* gene has been cloned and the gene product identified as a 24,000 m.w. protein (Little and Harper 1979) which represses its own expression (Brent and Ptashne 1980) as well as the expression of the *recA* gene (Little et al. 1981) and the *uvrA* and *uvrB* genes (Fogliano and Schendel 1981; Kenyon and Walker 1981) and may repress other unlinked *lexA* target genes. The *lexA* gene product acts as a repressor by binding to similar palindromeic sequences in the operator regions of these genes, excluding RNA polymerase from the promoter and thus blocking transcription (Brent and Ptashne 1981). Inactivation of the *lexA* protein by proteolytic cleavage results in increased synthesis of the *recA* protein as well as the derepression of other *din* (*d*amage-*in*ducible) genes (Kenyon and Walker 1980). The *recA* protease is inactive when the single-strand gaps, which act as a cofactor, are repaired. The *lexA* protein, which is not cleaved by the inactive *recA* protease, accumulates and represses the "SOS" functions thus returning the cell to its normal growth state.

The ability to recover from injuries, whether mechanical, chemical or radiation, is characteristic of all living things and favored by natural selection. Therefore, it is not surprising to find genetic systems which enable bacteria and their viruses to recover from the potentially lethal effects of ultraviolet light, nor is it surprising to find this repair system tightly regulated and subject to loss by mutation. Most of the premutational lesions in bacterial and viral DNA can be eliminated by essentially "error-proof" repair pathways. The inducible "SOS"-repair system, which is coordinately expresssed with a number of other phenomena, is efficient, but error-prone. The induction of a DNA polymerase that allows postreplication repair past noncoding lesions in the template DNA increases cell survival and mutations and thereby adds to the variability and survival of the species.

Acknowledgment. The author wishes to thank Dr. William L. Belser, University of California, Riverside, for critical discussions during the writing of this review.

References

Azizbekyan, R.R., Galitskaya, L.A.: On the mechanism of UV-reactivation of *Bacillus subtilis* phage. Sov. Genet. 9, 869-872 (1975)

Barbour, S.D., Clark, A.J.: Biochemical and genetic studies of recombination proficiency in *Escherichia coli*. I. Enzymatic activity associated with *recB* and *recC* genes. Proc. Natl. Acad. Sci. USA 65, 955-961 (1970)

Bernstein, C.: Deoxyribonucleic acid repair in bacteriophage. Microbiol. Rev. 45, 72-98 (1981)

Bertani, L.E.: Host-dependent induction of phage mutants and lysogenization. Virology 12, 553-569 (1960)

Blanco, M., Devoret, R.: Repair mechanisms involved in prophage reactivation and UV-reactivation of UV-irradiated phage λ. Mutat. Res. 17, 293-305 (1973)

Bonura, T., Smith, K.C.: Enzymatic production of deoxyribonucleic acid double-strand breaks after ultraviolet irradiation of *Escherichia coli* K-12. J. Bacteriol. 121, 511-517 (1975)

Borek, E., Ryan, A.: The transfer of irradiation-elicited induction in a lysogenic organism. Proc. Natl. Acad. Sci. USA 44, 374-377 (1958)
Boyce, R.P., Howard-Flanders, P.: Release of ultraviolet light-induced thymine dimers from DNA in *E. coli* K-12. Proc. Natl. Acad. Sci. USA 51, 293-300 (1964)
Boyle, J.M., Setlow, R.B.: Correlations between host-cell reactivation, ultraviolet reactivation and pyrimidine dimer excision in the DNA of phage λ. J. Mol. Biol. 51, 131-144 (1970)
Brent, R., Ptashne, M.: The *lexA* gene product represses its own promoter. Proc. Natl. Acad. Sci. USA 77, 1932-1936 (1980)
Brent, R., Ptashne, M.: Mechanism of action of the *lexA* gene product. Proc. Natl. Acad. Sci. USA 78, 4204-4208 (1981)
Bresler, S.E., Kalinin, V.L., Shelegedin, V.N.: W-reactivation and W-mutagenesis of gamma-irradiated phage lambda. Mutat. Res. 49, 341-355 (1978)
Brooks, K., Clark, A.J.: Behaviour of λ bacteriophage in a recombination deficient strain of *Escherichia coli*. J. Virol. 1, 283-293 (1967)
Caillet-Fauquet, P., Defais, M.: UV reactivation of phage in a *polA* mutant of *E. coli*. Mutat. Res. 15, 353-355 (1972)
Cassuto, E., West, S.C., Mursalim, J., Conlon, S., Howard-Flanders, P.: Initiation of genetic recombination: Homologous pairing between duplex DNA molecules promoted by recA protein. Proc. Natl. Acad. Sci. USA 77, 3962-3966 (1980)
Castellazzi, M., George, J., Buttin, G.: Prophage induction and cell division in *E. coli*. I. Further characterization of the thermosensitive mutation of *tif-I* whose expression mimics the effect of UV irradiation. Mol. Gen. Genet. 119, 139-152 (1972a)
Castellazzi, M., George, J., Buttin, G.: Prophage induction and cell division in *E. coli*. II. Linked (*recA*, *zab*) and unlinked (*lex*) suppressors of *tif-1*-mediated induction and filamentation. Mol. Gen. Genet. 119, 153-174 (1972b)
Clark, A.J., Margulies, A.D.: Isolation and characterization of recombination-deficient mutants of *Escherichia coli* K12. Proc. Natl. Acad. Sci. USA 53, 451-459 (1965)
Craig, N.L., Roberts, J.W.: *E. coli* recA protein-directed cleavage of phage λ repressor requires polynucleotide. Nature 283, 26-30 (1980)
DasGupta, C.K., Poddar, R.K.: Ultraviolet reactivation of the single stranded DNA phage ϕX174. Mol. Gen. Genet. 139, 77-91 (1975)
Defais, M., Fauquet, P., Radman, M., Errera, M.: Ultraviolet reactivation and ultraviolet mutagenesis of λ in different genetic systems. Virology 43, 495-503 (1971)
Devoret, R.: Repair mechanisms of radiation damage: a third repair process. Curr. Top. Radiat. Res. Q. 9, 11-14 (1973)
Devoret, R., Blanco, M., George, J., Radman, M.: Recovery of phage λ from ultraviolet damage. In: Molecular Mechanisms for Repair of DNA (eds. R.C. Hanawalt, R.B. Setlow), pp. 155-171. New York: Plenum Press 1975
Devoret, R., George, J.: Induction indirecte du prophage λ par le rayonnement ultraviolet. Mutat. Res. 4, 713-734 (1967)
Donch, J., Grenn, M.H.L., Greenberg, J.: Interaction of the *exr* and *lon* genes in *Escherichia coli*. J. Bacteriol. 96, 1704-1710 (1968)
Donch, J., Greenberg, J., Green, M.H.L.: Repression of induction by uv of λ phage by *exrA* mutations in *Escherichia coli*. Genet. Res. 15, 87-97 (1970).
Dulbecco, R.: Experiments on photoreactivation of bacteriophages inactivated with ultraviolet radiation. J. Bacteriol. 59, 329-347 (1950)
Echols, H., Gingery, R.: Mutants of bacteriophage λ defective in vegetative genetic recombination. J. Mol. Biol. 34, 239-249 (1968)
Fogliano, M., Schendel, P.F.: Evidence for the inducibility of the *uvrB* operon. Nature 289, 196-198 (1981)

Fuerst, C.R., Siminovitch, L.: Characterization of an unusual defective lysogenic strain of *Escherichia coli* K12 (λ). Virology 27, 449-451 (1965)
Garen, A., Zinder, N.D.: Radiological evidence for partial genetic homology between bacteriophage and host bacteria. Virology 1, 347-376 (1955)
George, J., Devoret, R.: Conjugal transfer of UV-damaged F-prime sex factors and indirect induction of prophage λ. Mol. Gen. Genet. 111, 103-119 (1971)
George, J., Devoret, R., Radman, M.: Indirect ultraviolet-reactivation of phage λ. Proc. Natl. Acad. Sci. USA 71, 144-147 (1974)
Gottesman, M.M., Hicks, M.L., Gellert, M.: Genetics and function of DNA ligase in *Escherichia coli*. J. Mol. Biol. 77, 531-547 (1973)
Green, M.H.L., Greenberg, J., Donch, J.: Effect of a *recA* gene on cell division and capsular polysaccharide production in a *lon* strain of *Escherichia coli*. Genet. Res. 14, 159-162 (1969)
Grossman, L., Braun, A., Feldberg, R., Mahler, I.: Enzymatic repair of DNA. Annu. Rev. Biochem. 44, 19-43 (1975)
Gudas, L.J., Mount, D.W.: Identification of the *recA* (*tif*) gene product of *Escherichia coli*. Proc. Natl. Acad. Sci. USA 74, 5280-5284 (1977)
Hanawalt, P.: Cellular recovery from photochemical damage. In: Photophysiology, Vol. 4 (ed. A.C. Giese), pp. 203-251. New York: Academic Press 1968
Hanawalt, P.C., Setlow, R.B. (eds.): Molecular Mechanisms for the Repair of DNA. New York: Plenum Press 1975
Harm, W.: On the relationship between host-cell reactivation and UV-reactivation in UV-inactivated phages. Z. Vererbungslehre 94, 67-79 (1963)
Harm, W.: Comment on the relationship between UV reactivation and host-cell reactivation in phage. Virology 29, 494 (1966)
Harm, W., Rupert, C.S.: Infection of transformable cells of *Haemophilus influenzae* by bacteriophage and bacteriophage DNA. Z. Vererbungslehre 94, 336-348 (1963)
Hart, M.G.R., Ellison, J.: Ultraviolet reactivation in bacteriophage lambda. J. Gen. Virol. 8, 197-208 (1970)
Hayes, W.: The Genetics of Bacteria and their Viruses, 2nd ed., pp. 328-338. Oxford: Blackwell 1974
Hertman, I., Luria, S.E.: Transduction studies on the role of a rec^+ gene in the ultraviolet induction of prophage lambda. J. Mol. Biol. 23, 117-133 (1967)
Horii, T., Ogawa, T., Ogawa, H.: Organization of the *recA* gene of *Escherichia coli*. Proc. Natl. Acad. Sci. USA 77, 313-317 (1980)
Howard-Flanders, P., Boyce, R.P.: DNA repair and genetic recombination: studies on mutants of *Escherichia coli* defective in these processes. Radiat. Res. 29 (Suppl. 6), 156-184 (1966)
Howard-Flanders, P., Theriot, L.: Mutants of *Escherichia coli* K-12 defective in DNA repair and in genetic recombination. Genetics 53, 1137-1150 (1966)
Howard-Flanders, P., Boyce, R.P., Theriot, L.: Three loci in *Escherichia coli* K-12 that control the excision of thymine dimers and certain other mutagen products from DNA. Genetics 53, 1119-1136 (1966)
Howard-Flanders, P., Rupp, W.D., Wilkins, B.M., Cole, R.S.: DNA replication and recombination after UV irradiation. Cold Spring Harbor Symp. Quant. Biol. 33, 195-205 (1968)
Howarth, S.: Resistance to the bactericidal effect of ultraviolet radiation conferred on Enterobacteria by the colicine factor *colI*. J. Gen. Microbiol. 40, 43-55 (1965)
Howarth, S.: Increase in frequency of ultraviolet-induced mutation brought about by the colicine factor *colI* in *Salmonella typhimurium*. Mutat. Res. 3, 129-134 (1966)
Ikeda, H., Tomizawa, J.: Prophage PI, an extrachromosomal replication unit. Cold Spring Harbor Symp. Quant. Biol. 33, 791-798 (1968)

Jacob, F., Wollman, E.L.: Induction of phage development in lysogenic bacteria. Cold Spring Harbor Symp. Quant. Biol. 18, 101-121 (1953)

Jacob, F., Wollman, E.L.: Lysogeny. In: The Viruses, Vol. 2 (eds. F.M. Burnet, W.M. Stanley), pp. 319-351. New York: Academic Press 1959

Kelner, A.: Photoreactivation of ultraviolet-irradiated *Escherichia coli* with special reference to the dose-reduction principle and to ultraviolet-induced mutations. J. Bacteriol. 58, 511-522 (1949)

Kenyon, C.J., Walker, G.C.: DNA-damaging agents stimulate gene expression at specific loci in *Escherichia coli*. Proc. Natl. Acad. Sci. USA 77, 2819-2823 (1980)

Kenyon, C.J., Walker, G.C.: Expression of the *E. coli uvrA* gene is inducible. Nature 289, 808-810 (1981)

Kerr, T.L., Hart, M.G.R.: Effects of the *rec* and *exr* mutations of *Escherichia coli* on UV reactivation of bacteriophage lambda damaged by different agents. Mutat. Res. 15, 247-258 (1972)

Kerr, T.L., Hart, M.G.R.: Effects of the *recA*, *lex* and *exrA* mutations on the survival of damaged λ and PI phages in lysogenic and non-lysogenic strains of *Escherichia coli* K12. Mutat. Res. 18, 113-116 (1973)

Kirby, E.P., Jacob, F., Goldthwait, D.A.: Prophage induction and filament formation in a mutant strain of *Escherichia coli*. Proc. Natl. Acad. Sci. USA 58, 1903-1910 (1967)

Kneser, H.: Relationship between K-reactivation and UV-reactivation of bacteriophage λ. Virology 36, 303-305 (1968)

Kneser, H., Metzger, K., Sauerbier, W.: Evidence of different mechanisms for ultraviolet reactivation and "ordinary host cell reactivation" of phage λ. Virology 27, 213-221 (1965)

Knudson, G.B.: A Study of Error-prone Repair in Species of *Serratia*. Ph. D. Thesis, University of California, Riverside (1977)

Korn, D., Weissbach, A.: Thymineless induction of *Escherichia coli* K12 (λ). Biochim. Biophys. Acta 61, 775-790 (1962)

Krishnapillai, V.: Resistance to ultraviolet light and enhanced mutagenesis conferred by *Pseudomonas aeruginosa* plasmids. Mutat. Res. 29, 363-372 (1975)

Kushner, S.R., Kaplan, J.C., Ono, H., Grossman, L.: Enzymatic repair of deoxyribonucleic acid. IV. Mechanism of photoproduct excision. Biochemistry 10, 3325-3334 (1971)

Lehrbach, P., Kung, A.H.C., Lee, B.T.O., Jacoby, G.A.: Plasmid modification of radiation and chemical-mutagen sensitivity in *Pseudomonas aeruginosa*. J. Gen. Microbiol. 98, 167-176 (1977)

Lehrbach, P.R., Kung, A.H.C., Lee, B.T.O.: R plasmids which alter ultraviolet light-sensitivity and enhance ultraviolet light-induced mutability in *Pseudomonas aeruginosa*. J. Gen. Microbiol. 108, 119-123 (1978)

Lewin, B.: Gene Expression, Vol. 1, pp. 495-523. London: Wiley 1974

Little, J.W., Harper, J.E.: Identification of the *lexA* gene product of *Escherichia coli* K-12. Proc. Natl. Acad. Sci. USA 76, 6147-6151 (1979)

Little, J.W., Edmiston, S.H., Pacelli, L.Z., Mount, D.W.: Cleavage of the *Escherichia coli lexA* protein by the *recA* protease. Proc. Natl. Acad. Sci. USA 77, 3225-3229 (1980)

Little, J.W., Mount, D.W., Yanisch-Perron, C.R.: Purified *lexA* protein is a repressor of the *recA* and *lexA* genes. Proc. Natl. Acad. Sci. USA 78, 4199-4203 (1981)

Luria, S.E.: Reactivation of ultraviolet-irradiated bacteriophage by multiple infection. J. Cell. Comp. Physiol. 39 (Suppl.), 119-123 (1952)

Lwoff, A., Siminovitch, L., Kjeldgaard, N.: Induction of the production of bacteriophages from lysogenic bacteria. Ann. Inst. Pasteur 79, 815-859 (1950)

MacPhee, D.G.: Effect of *rec* mutations on the ultraviolet protecting and mutation-enhancing properties of the plasmid R-Utrecht in *Salmonella typhimurium*. Mutat. Res. 19, 357-359 (1973)

Marmur, J., Grossman, L.: Ultraviolet light induced linking of deoxyribonucleic acid strands and its reversal by photoreactivating enzyme. Proc. Natl. Acad. Sci. USA 47, 778-787 (1961)

Mattern, I.E., Van Winden, M.P., Rörsch, A.: The range of action of genes controlling radiation sensitivity in *Escherichia coli*. Mutat. Res. 2, 111-131 (1965)

McCann, J., Spingarn, N.E., Kobori, J., Ames, B.N.: Detection of carbinogens and mutagens: bacterial tester strains with R factor plasmids. Proc. Natl. Acad. Sci. USA 72, 979-983 (1975)

McEntee, K., Weinstock, G.M., Lehman, I.R.: Initiation of general recombination catalyzed in vitro by the *recA* protein of *Escherichia coli*. Proc. Natl. Acad. Sci. USA 76, 2615-2619 (1979)

McKee, R.A., Hart, M.G.R.: Effects of the *Escherichia coli* K12 *recA56*, *uvrB* and *polA* mutations on UV reactivation in bacteriophage T7. Mutat. Res. 28, 305-308 (1975)

Miehl, R., Miller, M., Yasbin, R.E.: Plasmid mediated enhancement of uv resistance in *Streptococcus faecalis*. Plasmid 3, 128-134 (1980)

Miura, A., Tomizawa, J.: Studies on radiation-sensitive mutants of *E. coli*. III. Participation of the Rec system in induction of mutation by ultraviolet irradiation. Mol. Gen. Genet. 103, 1-10 (1968)

Monk, M.: Induction of phage λ by transferred irradiated *colI* DNA. Mol. Gen. Genet. 106, 14-24 (1969)

Monk, M., Kinross, J.: The kinetics of derepression of prophage λ following ultraviolet irradiation of lysogenic cells. Mol. Gen. Genet. 137, 263-268 (1975)

Monti-Bragadin, C., Babudri, N., Venturini, S.: Plasmid-borne error-prone DNA repair. In: DNA Synthesis (eds. I. Molineux, M. Kohiyama), pp. 1025-1032. New York: Plenum Press 1978

Morse, L.S., Pauling, C.: Induction of error-prone repair as a consequence of DNA ligase deficiency in *Escherichia coli*. Proc. Natl. Acad. Sci. USA 72, 4645-4649 (1975)

Mortelsman, K.E., Stocker, B.A.D.: Ultraviolet light protection, enhancement of ultraviolet light mutagenesis, and mutator effect of plasmic R46 in *Salmonella typhimurium*. J. Bacteriol. 128, 271-282 (1976)

Mount, D.W.: A mutant of *Escherichia coli* showing constitutive expression of the lysogenic induction and error-prone DNA repair pathways. Proc. Natl. Acad. Sci. USA 74, 300-304 (1977)

Mount, D.W., Low, K.B., Edmiston, S.J.: Dominant mutations (*lex*) in *Escherichia coli* K-12 which affect radiation sensitivity and frequency of ultraviolet light-induced mutations. J. Bacteriol. 112, 886-893 (1972)

Mount, D.W., Kosel, C.K., Walker, A.: Inducible, error-free DNA repair in *tsl recA* mutants of *E. coli*. Mol. Gen. Genet. 146, 37-41 (1976)

Ogawa, H.: Genetic locations of *uvrD* and *pol* genes of *E. coli*. Mol. Gen. Genet. 108, 378-381 (1970)

Ogawa, H., Tomizawa, J.: Ultraviolet reactivation of lambda phage: assay of infectivity of DNA molecules by spheroplast transfection. J. Mol. Biol. 73, 397-406 (1973)

Ogawa, H., Shimada, K., Tomizawa, J.: Studies on radiation-sensitive mutants of *E. coli*. I. Mutants defective in the repair synthesis. Mol. Gen. Genet. 101, 227-244 (1968)

Ono, J., Shimazu, Y.: Ultraviolet reactivation of a bacteriophage containing a single-stranded deoxyribonucleic acid as a genetic element. Virology 29, 295-302 (1966)

Otsuji, N., Okubo, S.: Reactivation of ultraviolet and nitrous acid-inactivated phages by host cells. Virology 12, 607-609 (1960)

Paterson, M.C., Boyle, J.M., Setlow, R.B.: Ultraviolet- and X-ray-induced responses of a deoxyribonucleic acid polymerase-deficient mutant of *Escherichia coli*. J. Bacteriol. 107, 61-67 (1971)

Pauling, C., Hamm, L.: Properties of a temperature-sensitive radiation-sensitive mutant of *Escherichia coli*. II. DNA replication. Proc. Natl. Acad. Sci. USA 64, 1195-1202 (1969)

Phizicky, E.M., Roberts, J.W.: Kinetics of *recA* protein-directed inactivation of repressors of phage λ and phage P22. J. Mol. Biol. 139, 319-328 (1980)

Ptashne, M.: Isolation of the λ phage repressor. Proc. Natl. Acad. Sci. USA 57, 306-313 (1967)

Radman, M.: Phenomenology of an inducible mutagenic DNA repair pathway in *Escherichia coli*: SOS repair hypothesis. In: Molecular and Environmental Aspects of Mutagenesis (ed. M.C. Miller), pp. 128-142. Springfield, Ill.: Thomas 1974

Radman, M., Devoret, R.: UV-reactivation of bacteriophage λ in excision repair-deficient hosts: independence of red functions and attachment regions. Virology 43, 504-506 (1971)

Roberts, J.W., Roberts, C.W., Craig, N.L.: *Escherichia coli recA* gene product inactivates phage λ repressor. Proc. Natl. Acad. Sci. USA 75, 4714-4718 (1978)

Rosner, J.L., Kass, L.R., Yarmolinsky, M.B.: Parallel behavior of F and PI in causing indirect induction of lysogenic bacteria. Cold Spring Harbor Symp. Quant. Biol. 33, 758-789 (1968)

Rothman, R.H., Margossian, L.J., Clark, A.J.: W-reactivation of phage lambda in *recF*, *recL*, *uvrA* and *uvrB* mutants of *E. coli* K-12. Mol. Gen. Genet. 169, 279-287 (1979)

Roulland-Dussoix, D.: Dégradation par la cellule hôte du DNA du bactériophage lambda irradié par le rayonnement ultraviolet. Mutat. Res. 4, 241-252 (1967)

Rupert, C.S.: Repair of ultraviolet damage in cellular DNA. J. Cell. Comp. Physiol. 58 (Suppl.), 57-68 (1961)

Rupert, C.S., Harm, W.: Reactivation after photobiological damage. Adv. Radiat. Biol. 2, 1-81 (1966)

Rupp, W.D., Howard-Flanders, P.: Discontinuities in the DNA synthesized in an excision-defective strain of *Escherichia coli* following ultraviolet irradiation. J. Mol. Biol. 31, 291-304 (1968)

Rupp, W.D., Wilde, C.E., III, Reno, D., Howard-Flanders, P.: Exchanges between DNA strands in ultraviolet-irradiated *Escherichia coli*. J. Mol. Biol. 61, 25-44 (1971)

Sancar, A., Stachelek, C., Konigsberg, W., Rupp, W.D.: Sequences of the *recA* gene and protein. Proc. Natl. Acad. Sci. USA 77, 2611-2615 (1980)

Sedgwick, S.G.: Evidence for an inducible error prone repair system in *Escherichia coli*. Biophys. J. 15, 301a (1975a)

Sedgwick, S.G.: Inducible error-prone repair in *Escherichia coli*. Proc. Natl. Acad. Sci. USA 72, 2753-2757 (1975b)

Setlow, J.K., Setlow, R.B.: Nature of the photoreactivable ultraviolet lesion in deoxyribonucleic acid. Nature 197, 560-562 (1963)

Setlow, R.B., Carrier, W.L.: The disappearance of thymine dimers from DNA: an error-correcting mechanism. Proc. Natl. Acad. Sci. USA 51, 226-231 (1964)

Setlow, R.B., Carrier, W.L.: Pyrimidine dimers in ultraviolet-irradiated DNA's. J. Mol. Biol. 17, 237-254 (1966)

Shibata, T., DasGupta, C., Cunningham, R.P., Radding, C.M.: Purified *Escherichia coli recA* protein catalyzes homologous pairing of superhelical DNA and single-stranded fragments. Proc. Natl. Acad. Sci. USA 76, 1638-1642 (1979)

Shimada, K., Ogawa, H., Tomizawa, J.: Studies on radiation-sensitive mutants of *E. coli*. II. Breakage and repair of ultraviolet irradiated intracellular DNA of phage lambda. Mol. Gen. Genet. 101, 245-256 (1968)

Siccardi, A.G.: Effect of R factors and other plasmids in ultraviolet susceptibility and host cell reactivation property of *Escherichia coli*. J. Bacteriol. 100, 337-346 (1969)
Stent, G.S.: Mating in the reproduction of bacterial viruses. Adv. Virus Res. 5, 95-149 (1958)
Tessman, E.S.: Growth and mutation of phage T1 on ultraviolet-irradiated host cells. Virology 2, 679-688 (1956)
Tessman, E.S., Ozaki, T.: The interaction of phage S13 with ultraviolet-irradiated host cells and properties of the ultraviolet-irradiated phage. Virology 12, 431-449 (1960)
Tomilin, N.V., Mosevitskaya, T.V.: Ultraviolet reactivation and ultraviolet mutagenesis of infectious lambda DNA: strong inhibition by treatment of DNA in vitro with UV-endonuclease from *Micrococcus luteus*. Mutat. Res. 27, 147-156 (1975)
Tomizawa, J., Ogawa, T.: Effect of ultraviolet irradiation on bacteriophage lambda immunity. J. Mol. Biol. 23, 247-263 (1967)
Van de Putte, P., Van Sluis, C.A., Van Dillewijn, J., Rörsch, A.: The location of genes controlling radiation sensitivity in *Escherichia coli*. Mutat. Res. 2, 97-110 (1965)
Vizdalova, M.: The inactivating effect of hydroxylamine on *E. coli* phages and the possibility of repair of the resultant damage. Int. J. Radiat. Oncol. Biol. Phys. 16, 147-155 (1969)
Wackernagel, W., Winkler, U.: A mutation in *Escherichia coli* enhancing the UV-mtability of phage λ but not of its infectious DNA in a spheroplast assay. Mol. Gen. Genet. 114, 68-79 (1971)
Waleh, N.S., Stocker, B.A.D.: Effect of host *lex*, *recA*, *recF*, and *uvrD* genotypes on the ultraviolet light-protecting and related properties of plasmid R46 in *Escherichia coli*. J. Bacteriol. 137, 830-838 (1979)
Walker, G.C.: Plasmid (pKM101)-mediated enhancement of repair and mutagenesis: dependence on chromosomal genes in *Escherichia coli* K-12. Mol. Gen. Genet. 152, 93-103 (1977)
Walker, G.C.: Inducible reactivation and mutagenesis of UV-irradiated bacteriophage P22 in *Salmonella typhimurium* LT2 containing the plasmid pKM101. J. Bacteriol. 135, 415-421 (1978)
Walker, G.C., Dobson, P.P.: Mutagenesis and repair deficiencies of *Escherichia coli umuC* mutants are suppressed by the plasmid pKM101. Mol. Gen. Genet. 172, 17-24 (1979)
Weigle, J.J.: Induction of mutations in a bacterial virus. Proc. Natl. Acad. Sci. USA 39, 628-636 (1953)
Weigle, J.J.: Story and structure of the λ transducing phage. In: Phage and the Origins of Molecular Biology (eds. J. Cairns, G.S. Stent, J.D. Watson), pp. 123-138. Cold Spring Harbor, NY: Cold Spring Harbor Laboratory 1966
Weigle, J.J., Dulbecco, R.: Induction of mutations in bacteriophage T3 by ultra-violet light. Experientia 9, 372-373 (1953)
West, S.C., Cassuto, E., Howard-Flanders, P.: recA protein promotes homologous-pairing and strand-exchange reactions between duplex DNA molecules. Proc. Natl. Acad. Sci. USA 78, 2100-2104 (1981)
Witkin, E.M.: Time, temperature, and protein synthesis: a study of ultraviolet-induced mutation in bacteria. Cold Spring Harbor Symp. Quant. Biol. 21, 123-138 (1956)
Witkin, E.M.: Mutation-proof and mutation-prone modes of survival in derivatives of *Escherichia coli* B differing in sensitivity to ultraviolet light. Brookhaven Symp. Biol. 20, 17-55 (1967a)
Witkin, E.M.: The radiation sensitivity of *Escherichia coli* B: a hypothesis relating filament formation and prophage induction. Proc. Natl. Acad. Sci. USA 57, 1275-1279 (1967b)

Witkin, E.M.: The role of DNA repair and recombination in mutagenesis. Proc. 12th Int. Congr. Genet. 3, 225-245 (1969a)
Witkin, E.M.: Ultraviolet-induced mutation and DNA repair. Annu. Rev. Genet. 3, 525-552 (1969b)
Witkin, E.M.: Ultraviolet-induced mutation and DNA repair. Annu. Rev. Microbiol. 23, 487-514 (1969c)
Witkin, E.M.: Thermal enhancement of ultraviolet mutability in a *tif-1 uvrA* derivative of *Escherichia coli* B/r: evidence that ultraviolet mutagenesis depends upon an inducible function. Proc. Natl. Acad. Sci. USA 71, 1930-1934 (1974)
Witkin, E.M.: Elevated mutability of *polA* and *uvrA polA* derivatives of *Escherichia coli* B/r at sublethal doses of ultraviolet light: evidence for an inducible error-prone repair system ("SOS repair") and its anomalous expression in these strains. Genetics 79, 199-213 (1975)
Witkin, E.M.: Ultraviolet mutagenesis and inducible DNA repair in *Escherichia coli*. Bacteriol. Rev. 40, 869-907 (1976)
Witkin, E.M., George, D.L.: Ultraviolet mutagenesis in *polA* and *uvrA polA* derivatives of *Escherichia coli* B/r: evidence for an inducible error-prone repair system. Genetics 73 (Suppl.), 91-110 (1973)
Wulff, D.L., Rupert, C.S.: Disappearance of the thymine photodimer in ultraviolet irradiated DNA upon treatment with a photoreactivating enzmye from baker's yeast. Biochem. Biophys. Res. Commun. 7, 237-240 (1962)
Yamamoto, N.: Recombination: damage and repair of bacteriophage genome. Biochem. Biophys. Res. Commun. 27, 263-269 (1967)

Biological and Biochemical Actions of Trichothecene Mycotoxins

James R. Bamburg

A. Introduction

The trichothecenes are a structurally related group of sesquiterpenoid compounds produced by several genera and species of imperfect fungi. Most compounds in this category have a spiro epoxy group on carbons 12 and 13, although one trichothecene recently isolated and characterized lacks this functionality. The general structural formula and the numbering system for the trichothecene ring system are shown in Fig. 1. The group of compounds take their name (Godtfredsen et al. 1967) from the first isolate, trichothecin, discovered in 1948 by Freeman and Morrison as a result of screening cultures of *Trichothecium roseum* for metabolites with antifungal activity. Since that time over sixty trichothecenes have been isolated and characterized. These trichothecenes were discovered as a result of their unusual biological activities which were detected as a result of screening fungi for antifungal agents, antibiotics, phytotoxins, antileukemic agents, cytotoxins, and animal toxins (mycotoxins). Trichothecenes have been implicated as causative agents in a wide variety of animal and human health problems, some of which have resulted in the deaths of hundreds of thousands of people. These compounds have been the subject of many reviews, the first major one addressing their role in animal and human mycotoxicoses appearing in 1971 (Bamburg) and Strong, 1971). Since 1971 major reviews about trichothecenes have appeared every few years (Smalley and Strong 1974; Bamburg 1976; Ueno 1977; Doyle and Bradner 1980; Ueno 1980a).

Yet another review of the trichothecene literature might have seemed inappropriate at this time if it had not been for a single political event with world-wide repercussions which took place in September 1981. Based on the analysis of a single leaf and stem sample obtained from a region of Kampuchea where government forces reportedly attacked suspected enemy positions with chemicals referred to as "yellow rain" (Seagrave 1981), the State Department of the United States released a fact sheet which indirectly accused the governments of the Soviet Union, Vietnam and Laos of engaging in chemical warfare, employing trichothecene mycotoxins as the chemical agents in question. The "fact sheet" released September 14, 1982, stated in part:

> "Analysis of a leaf and stem sample from Kampuchea has revealed high levels of mycotoxins of the trichothecene group. The levels detected were up to twenty times greater than any recorded natural outbreak. Since normal background levels of these toxins are essentially undetectable, the high levels found are considered to be abnormal, and it is highly unlikely

Progress in Molecular and Subcellular Biology, Vol. 8, edited by F.E. Hahn

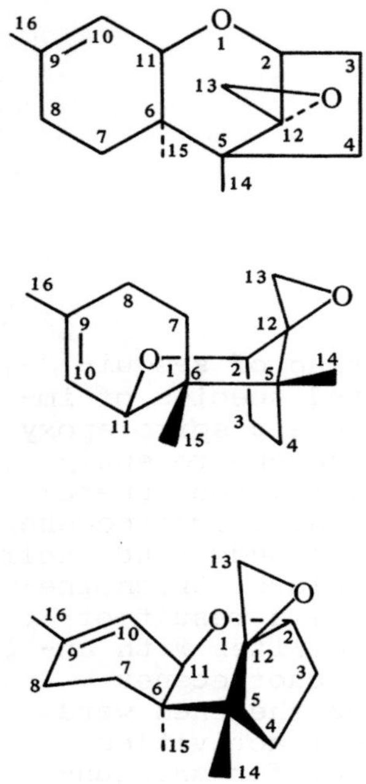

Fig. 1. Three representations of the chemical structure of trichothecene (12, 13-epoxytrichothec-9-ene)

> that such levels could have occurred in a natural intoxication. In point of fact, these mycotoxins do not occur naturally in Southeast Asia.
>
> Symptoms associated with these three lethal toxins specifically include rapid onet of vomiting, multiple hemorrhages of mucous membranes, bloody diarrhea, and severe itching and tingling of skin with formation of multiple small blisters, and death."

This announcement sparked a great deal of debate between the government and scientists in this country who questioned many aspects of the evidence. How reliable were the methods of analysis? How was the sample stored and transported (i.e., could the contamination with trichothecenes have occurred after collection of the plant?). Do trichothecene producing fungi occur in Southeast Asia and do they produce trichothecenes? What levels of trichothecenes are produced in natural outbreaks? Are the reported symptoms consistent with those of exposure to trichothecenes? By what mechanisms do trichothecenes cause toxicity? While not all of these questions can be answered specifically, this review of the trichothecenes will focus on their toxicological, biological and biochemical action and will address many of the questions related to trichothecene production, distribution and detection.

B. Trichothecene Structure and Genesis

I. Structure of Trichothecenes

The chemical structures and stereochemistry of all the trichothecenes have been confirmed by interconversion to a known derivative which was itself related either to the structure of trichodermol, determined by X-ray crystallography of its p-bromobenzoate ester (Abrahamsson and Nilsson 1964; Sigg et al. 1965; Abrahamsson and Nilsson 1966) or to verrucarin A whose structure was also confirmed by X-ray crystallography (McPhail and Simm 1966). By 1971, the time that the trichothecene literature was first reviewed, 22 naturally occurring members of this class had been discovered (Bamburg and Strong 1971). By 1975, that number had grown to 32 (Bamburg 1976) and by 1980, 48 naturally occurring trichothecenes had been reported (Ueno 1980a). At the time of writing, 68 trichothecenes have been isolated and characterized. For the most part, these compounds have been isolated from pure fungal cultures and not from "naturally infected" foodstuffs.

The structural features of the compounds allow them to be categorized in four groups to which Ueno (1977) assigned the designations of class A for the hydroxy and acyloxy substituted trichothecenes (Fig. 2), class B for the 8-keto derivatives (Fig. 3), class C for the macrocyclic derivatives (Fig. 4) and class D for the 7, 8 epoxy derivatives (Fig. 5). As in the previous reviews, the structures of the compounds will be reported in the 4 groups or classes which share certain features. In addition, a fifth group (class E, Fig. 6) needs to be established for the trichothecadiene, verrucarin K. More than likely, many of the more recently reported trichothecenes such as the trichoverrins, trichoverrols, and trichodermadienes serve as intermediates in the biosynthetic pathway to the macrocyclic trichothecenes (Jarvis et al. 1981b). Verrucarin K, on the other hand, may represent a branch point in the biosynthetic pathway (Fig. 7) or it might arise through a reduction of the 12, 13-epoxy group at a later stage. Structures of several new macrocyclic trichothecenes have been reported recently. Some of these have come from *Stachybotrys atra* (referred to as Satratoxins) and they are very similar or identical to the verrucarins and roridins which were originally isolated from *Myrothecium* spp.

II. Chemical Synthesis

The complete chemical synthesis of the trichothecene trichodermin was achieved by Colvin et al. in 1971. This synthetic route suffers from the disadvantage of a very low yield and it did not work in the synthesis of verrucarol (Colvin et al. 1978). Several new approaches to the synthesis of the trichothecene ring system in which a wider variety of substituents can be introduced have recently been reported (White et al. 1981; Pearson and Ong 1981). One of these methods, which uses organoiron complexes, has resulted in the stereoselective synthesis of 12,

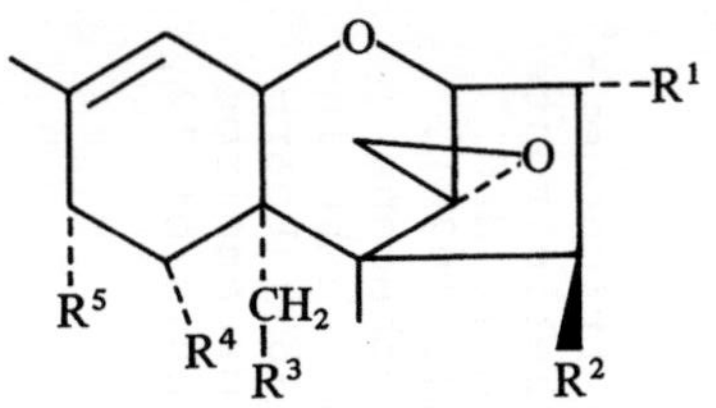

Fig. 2

Name	References[a]	R^1	R^2	R^3	R^4	R^5
Trichothecene	[1]	H	H	H	H	H
Trichodermol	[2,3]	H	OH	H	H	H
Trichodermin	[3]	H	OAC	H	H	H
Dihydroxytrichothecene	[1]	H	OH	H	H	OH
Verrucarol	[4]	H	OH	OH	H	H
Diacetylverrucarol	[5]	H	OAC	OAC	H	H
Scirpentriol	[6]	OH	OH	OH	H	H
T-2 Tetraol	[7]	OH	OH	OH	H	OH
Pentahydroxyscirpene	[8]	OH	OH	OH	OH	OH
15-Acetoxyscirpendiol	[9,10]	OH	OH	OAC	H	H
Diacetoxyscirpenol	[6,11,12]	OH	OAC	OAC	H	H
4-Acetoxyscirpendiol	[13,14]	OH	OAC	OH	H	H
Calonetrin	[15]	OAC	H	OAC	H	H
15-Desacetylcalonectrin	[15]	OAC	H	OH	H	H
Triacetoxyscripene	[6]	OAC	OAC	OAC	H	H
F. scirpi triacetate	[8]	OH	OAC	OAC	OH	OAC
7-Hydroxydiacetoxy-scirpenol	[16]	OH	OAC	OAC	OH	H

Fig. 2 (cont.)

Name	Reference[a]	R^1	R^2	R^3	R^4	R^5
Neosolaniol	[17]	OH	OAC	OAC	H	OH
7,8-Dihydroxydiacetoxy-scirpenol	[14]	OH	OAC	OAC	OH	OH
8-Acetylneosolaniol	[13,18]	OH	OAC	OAC	H	OAC
4,8-Diacetoxy-3,14-dihydroxy-trichothecene (NT-1)	[19,20]	OH	OAC	OH	H	OAC
4-Acetoxy-3,8,15-trihydroxy-trichothecene (NT-2)	[20]	OH	OAC	OH	H	OH
HT-2 Toxin	[21]	OH	OH	OAC	H	$OOCCH_2CH(CH_3)_2$
T-2 Toxin	[7]	OH	OAC	OAC	H	$OOCCH_2CH(CH_3)_2$
Acetyl T-2 Toxin	[22,23]	OAC	OAC	OAC	H	$OOCCH_2CH(CH_3)_2$
4-Desacetylneosolaniol	[24]	OH	OH	OAC	H	OH
3,4-Dihydroxy-15-acetoxy-8-(3-hydroxy-3-methylbutyryloxy) trichothecene	[24]	OH	OH	OAC	H	$OOCCH_2COH(CH_3)_2$
3,4,15-Trihydroxy-8-(3-hydroxy-3-methylbutyryloxy) trichothecene	[24]	OH	OH	OH	H	$OOCCH_2COH(CH_3)_2$
Trichodermadiene	[25]	H	$OOCCH{=}CHCH{=}CH\overset{O}{\widehat{CHCH}}CH_3$ (c) (t)(R)(R)	H	H	H
Trichodermadienol A	[26]	H	$OOCCH{=}CHCH{=}CHCHOHCHOHCH_3$ (c) (t)(S) (S)	H	H	H
Trichodermadienol B	[26]	H	$OOCCH{=}CHCH{=}CHCHOHCHOHCH_3$ (c) (t)(S) (R)	H	H	H
Trichoverrol A	[26]	H	$OOCCH{=}CHCH{=}CHCHOHCHOHCH_3$ (c) (t)(S) (S)	OH	H	H

Fig. 2 (cont.)

Name	Reference	R^1	R^2	R^3	R^4	R^5
Trichoverrol B	[26]	H	$OOCCH{=}CHCH{=}CHCHOHCHOHCH_3$ (c) (t) (S) (R)	OH	H	H
Trichoverrin A	[26]	H	$OOCCH{=}CHCH{=}CHCHOHCHOHCH_3$ (c) (t) (S) (S)	$OOCCH{=}C(CH_3)CH_2CH_2OH$ (E)	H	H
Trichoverrin B	[26]	H	$OOCCH{=}CHCH{=}CHCHOHCHOHCH_3$ (c) (t) (S) (R)	$OOCCH{=}C(CH_3)CH_2CH_2OH$ (E)	H	H

a
[1] Machida and Nozoe 1972; [2] Härri et al. 1962; [3] Godtfredsen and Vangedal 1964; [4] Gutzwiller et al. 1964; [5] Okuchi et al. 1968; [6] Sigg et al. 1965; [7] Bamburg et al. 1968a; [8] Grove 1970a; [9] Löffler et al. 1965; [10] Pathre et al. 1976; [11] Flury et al. 1965; [12] Dawkins et al. 1965; [13] Ishii et al. 1978; [14] Steyn et al. 1978; [15] Gardner et al. 1972; [16] Ishii 1975; [17] Ueno et al. 1972a; [18] Lansden et al. 1978; [19] Ilus et al. 1977; [20] Ishii and Ueno 1981; [21] Bamburg and Strong 1969; [22] Yoshizawa and Morooka 1973; [23] Kotsonis et al. 1975b; [24] Cole et al. 1981; [25] Jarvis et al. 1980b; [26] Jarvis et al. 1981b

Fig. 2. The structures of the trichothecenes with simple hydroxyl or acyl substituents (class A). Stereochemistry of the double bonds (c=cis, t=trans, Z or E) and the chiral centers (S or R) is shown for some of the structures)

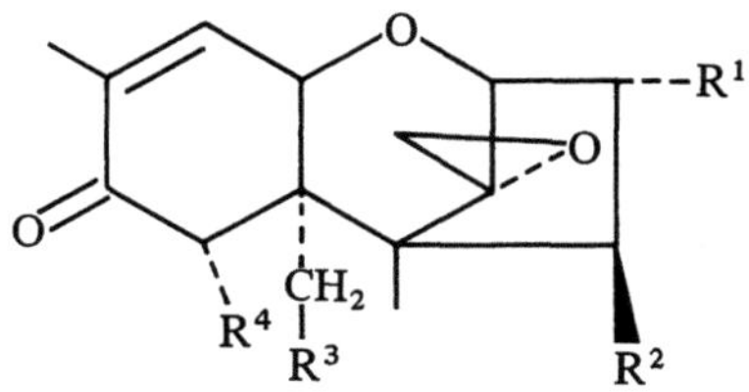

Fig. 3

Name	References[a]	R^1	R^2	R^3	R^4
Trichothecolone	[1,2]	H	OH	H	H
Trichothecin	[3,4,5]	H	$OOCCH{=}CHCH_3$	H	H
4-Acetyltrichothecolone	[6]	H	OAC	H	H
Nivalenol	[7,8]	OH	OH	OH	OH
Fusarenon-X	[8,9]	OH	OAC	OH	OH
4,15-Diacetylnivalenol	[8,10,11]	OH	OAC	OAC	OH
4-Desoxynivalenol (vomitoxin)	[12,13]	OH	H	OH	OH
3-Acetyldesoxynivalenol	[11,13]	OAC	H	OH	OH
7-Acetyldesoxynivalenol	[14]	OH	H	OH	OAC
3,15-Diacetyldesoxynivalenol	[14]	OAC	H	OAC	OH
4,7-Didesoxynivalenol	[15]	OH	H	OH	H
7-Desoxynivalenol	[16]	OH	OH	OH	H

[a][1] Achilladelis and Hanson 1969; [2] Freeman and Gill 1950; [3] Freeman and Morrison 1948; [4] Freeman and Morrison 1949; [5] Godtfredsen and Vangedal 1964; [6] Ghosal et al. 1982; [7] Tatsuno et al. 1968; [8] Grove 1970b; [9] Ueno et al. 1969; [10] Tidd 1967; [11] Blight and Grove 1974; [12] Vesonder et al. 1973; [13] Yoshizawa and Morooka 1973; [14] Yoshizawa and Morooka 1977; [15] Bennett et al. 1981; [16] Vesonder et al. 1976

Fig. 3. The structures of the 8-keto trichothecenes (class B)

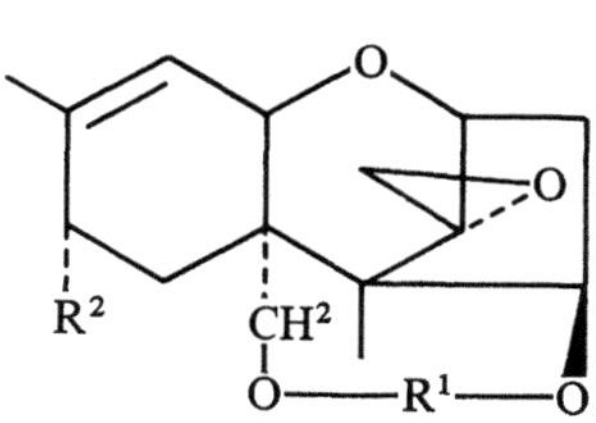

Fig. 4

Name	References[a]	R^1	R^2
Verrucarin A	[1,2]	$-\overset{O}{\overset{\parallel}{C}}CHOHCH(CH_3)CH_2CH_2O\overset{O}{\overset{\parallel}{C}}CH{=}CHCH{=}CH\overset{O}{\overset{\parallel}{C}}-$ (s) (R) (t) (c)	H

Fig. 4 (cont.)

Name	References[a]	R^1	R^2
2'-Dehydroverrucarin A	[3]	OO O O ‖‖ ‖ ‖ -CCCH(CH$_3$)CH$_2$CH$_2$OCCH=CHCH=CHC- (R) (t) (c)	H
Verrucarin B	[4]	O O O ‖(s)(R) ‖ ‖ -CCHC(CH$_3$)CH$_2$CH$_2$OCCH=CHCH=CHC- \/ (t) (c) O	
Verrucarin J (Satratoxin C)	[5]	O O O ‖ ‖ ‖ -CCH=C(CH$_3$)CH$_2$CH$_2$OCCH=CHCH=CHC- (E) (t) (c)	H
Verrucarin L	[6]	O O O ‖ ‖ ‖ -CCH=C(CH$_3$)CH$_2$CH$_2$OCCH=CHCH=CHC- (E) (t) (c)	OH
Verrucarin L acetate	[6]	O O O ‖ ‖ ‖ -CCH=C(CH$_3$)CH$_2$CH$_2$OCCH=CHCH=CHC- (E) (t) (c)	OAc
Roridin A	[7,18]	O O ‖ ‖ -CCHOHCH(CH$_3$)CH$_2$CH$_2$OCHCH=CHCH=CHC- (s) (R) | (t) (c) (R) CHOH | CH$_3$	H
Roridin D	[8,18]	O O ‖(s)(R) ‖ -CCHC(CH$_3$)CH$_2$CH$_2$OCHCH=CHCH=CHC- \/ | (t) (c) O CHOH | CH$_3$	H
Roridin E	[3,9,10]	O O ‖ (R) (t) (c) ‖ -CCH=C(CH$_3$)CH$_2$CH$_2$OCHCH=CHCH=CHC- (E) | (R) CHOH | CH$_3$	H
Roridin H	[11]	O O ‖ ‖ -CCH=C(CH$_3$)CH$_2$CHOCHCH=CHCH=CHC- (E) | | (t) (c) O — CH | CH$_3$	H

Fig. 4 (cont.)

Name	References[a]	R^1	R^2
Roridin J	[12]	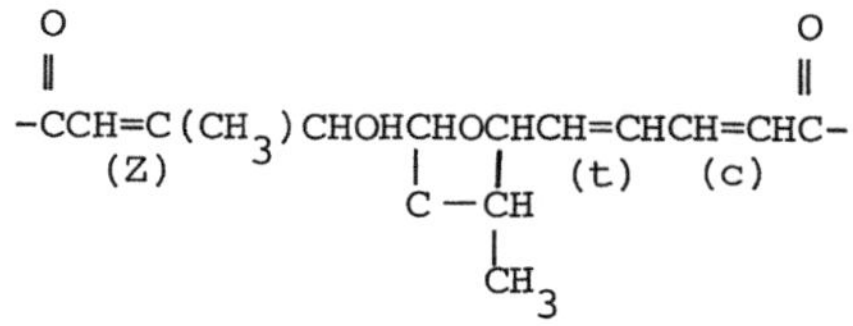	H
Roridin K	[13]	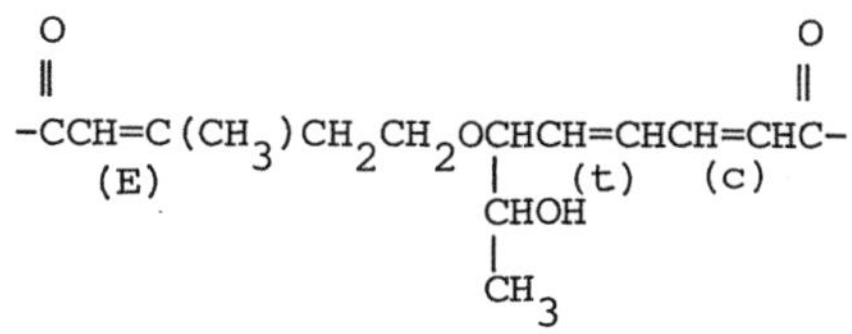	OAc
Satratoxin F	[14]	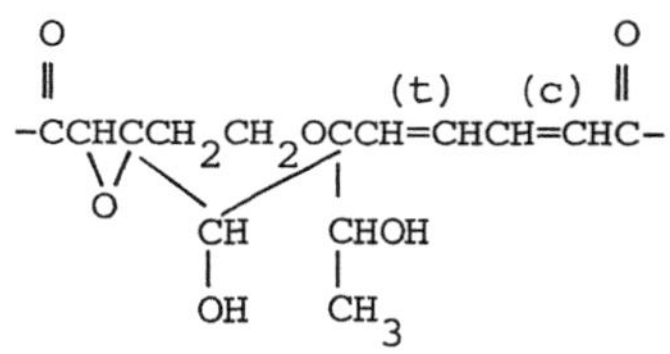	H
Satratoxin G	[14]	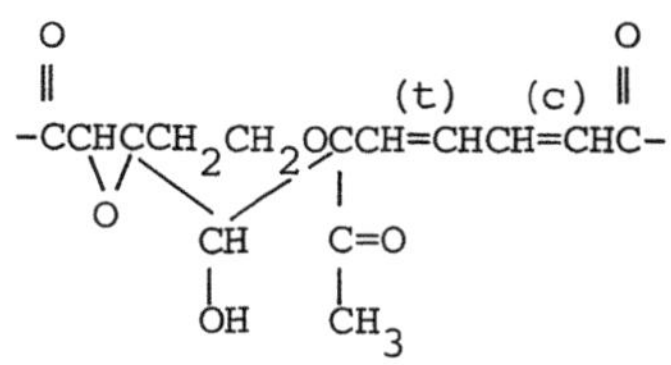	H
Satratoxin H	[15]	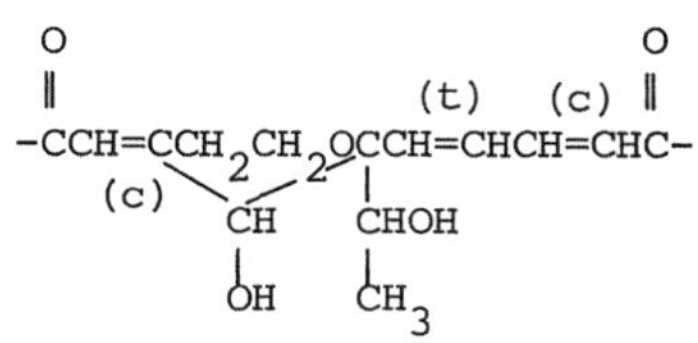	H
Vertisporin	[16]	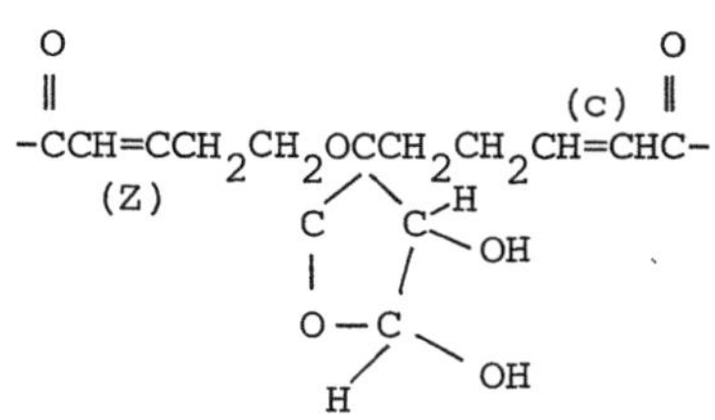	H
Isororidin E	[17,18]	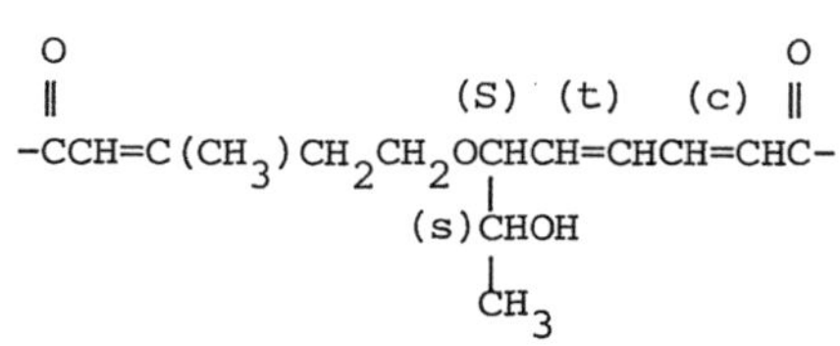	H

Fig. 4 (cont.)

a[1]Gutzwiller and Tamm 1965a; [2] McPhail and Sim 1966; [3] Zürcher and Tamm 1966; [4] Gutzwiller and Tamm 1965b; [5] Fetz et al. 1965; [6] Jarvis et al. 1981a; [7] Böhner and Tamm 1966b; [9] Eppley and Bailey 1973; [10] Traxler et al. 1970; [11] Traxler and Tamm 1970; [12] Jarvis et al. 1980a; [13] Midiwo 1981; [14] Eppley et al. 1980; [15] Eppley et al. 1977; [16] Minato et al. 1975; [17] Matsumoto et al. 1977; [18] Jarvis et al. 1982

Fig. 4. The structures of the macrocyclic trichothecenes (class C). Stereochemistry of the double bonds (*c*=cis, *t*=trans, Z or E) and the chiral centers (*S* or *R*) is shown for some of the structures

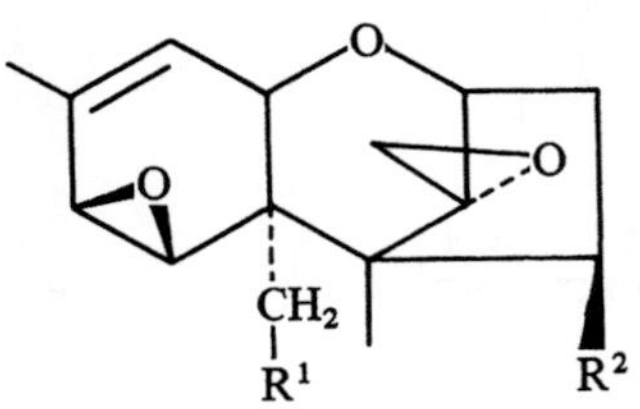

Fig. 5

Name	References[a]	R^1	R^2
Crotocol	[1]	H	OH
Crotocin	[1]	H	OOCCH=CHCH$_3$
		R^1-R^2	
7,8-Epoxyisororidin E	[2]	O ‖ -OCCH=C(CH$_3$)CH$_2$CH$_2$OCHCH=CHCH=CHC- (O ‖ on terminal C) | CHOH | CH$_3$	
7,8-Epoxyroridin H	[2]	O ‖ -OCCH=C(CH$_3$)CH$_2$CHOCHCH=CHCH=CHC- (O ‖ on terminal C) | | O — CH | CH$_3$	
Diepoxyroridin H	[2]	O ‖ -OCC-C(CH$_3$)CH$_2$CHOCHCH=CHCH=CHC- (O ‖ on terminal C) \/ O | | O — CH | CH$_3$	

a[1] Gyimesi and Melera 1967; [2] Matsumoto et al. 1977

Fig. 5. The structures of the 7,8-epoxy trichothecenes (class D)

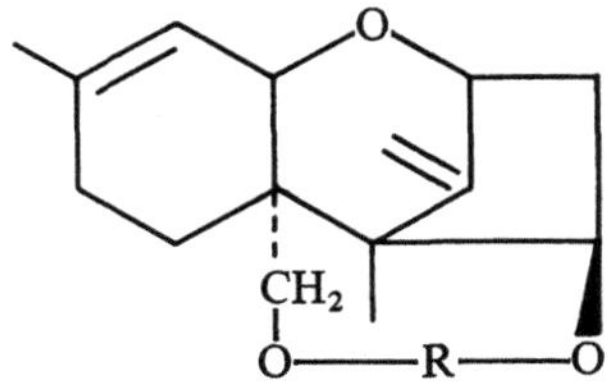

Name	Reference[a]	R
Verrucarin K	[1]	$-\overset{O}{\overset{\|}{C}}CHOHCH(CH_3)CH_2CH_2O\overset{O}{\overset{\|}{C}}CH{=}CHCH{=}CH\overset{O}{\overset{\|}{C}}-$

[a] [1] Breitenstein and Tamm 1977

Fig. 6. The structure of verrucarin K, a macrocyclic trichodermadiene

13-epoxy-14-methoxytrichothecene, the first reported 14-substituted trichothecene (Pearson and Ong 1981). The biological properties of this interesting trichothecene derivative have not yet been reported.

III. Biosynthesis

Jones and Lowe (1960) correctly deduced that 3 moles of mevalonic acid were incorporated into the trichothecene nucleus, thus recognizing the sesquiterpenoid nature of these compounds before their correct chemical structure was known. Many studies on the biosynthesis of these compounds have been adequately reviewed elsewhere (Bamburg 1976; Tamm 1977; Ciegler 1979). Farnesyl pyrophosphate cyclizes to form a trichodiene intermediate which may then form trichodiol or a dihydroxytrichodiene intermediate on the way to forming trichothecene or trichotheca-9,12-diene (Fig. 7). This latter compound is not thought to be a direct intermediate in trichothecene biosynthesis. However the discovery that verrucarin K (Fig. 7) lacked the 12,13-epoxy group (Breitenstein and Tamm 1977) may indicate that a branch point in the ring closing pathway occurs at trichodiene.

IV. Transformation of Trichothecenes in Plants

In the course of a continuing search for tumor inhibitors obtained from extracts of higher plants, Kupchan et al. (1976) discovered that trichothecenes were present in the extracts of a Brazilian shrub *Baccharis megapotamica*. This indeed was a surprising discovery since trichothecenes were only known to be secondary metabolites of a variety of *Fungi imperfecti*. Four novel trichothecenes were isolated, all macrocyclic compounds related to, but distinct from, roridin A. The structures of these compounds called baccharin, baccharinol, isobaccharin and isobaccharinol were determined by X-ray crystallography (Kupchan et al. 1977) and are shown in Fig. 8.

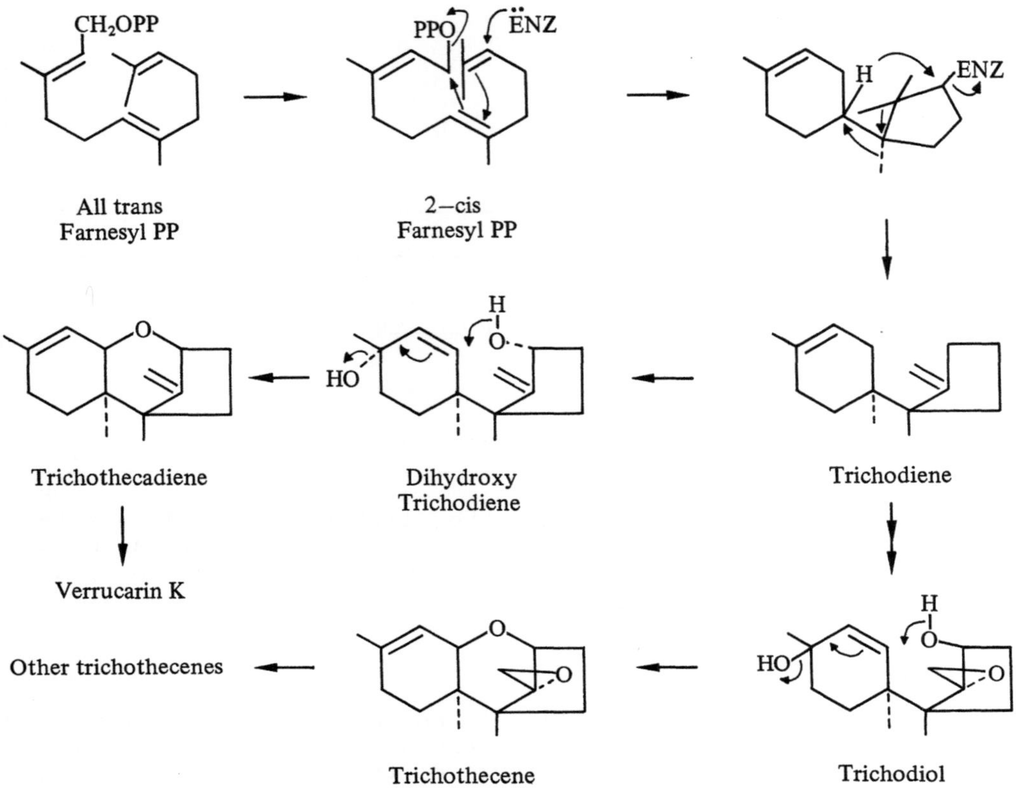

Fig. 7. Proposed biosynthetic pathway for the formation of trichothecenes from all trans farnesyl pyrophosphate. A branch point at trichodiene has been introduced to explain the formation of the trichothecadiene, verrucarin K. Data are from Machida and Nozoe 1972 and Breitenstein and Tamm 1977

That these compounds actually represent higher plant transformation products of the roridins was demonstrated by Jarvis et al. (1981c). These workers showed that *B. megapotamica*, grown from seed in non-Brazilian soil, does not contain these baccharinoids. When either roridin A or verrucarin A was added to the seedlings of *B. megapotamica*, both were rapidly taken up by the root system and translocated to the upper plant. In addition, both compounds were efficiently metabolized in high yield to their 8β-hydroxy derivatives with a slower epimerization reaction occurring at the C-2' position of the macrocyclic ring. Similar studies done with tomato, pepper and artichoke seedlings showed that roridin A was absorbed, translocated and, in the case of pepper and artichoke plants, also converted into the 8β-hydroxy derivative. However, these plants were severely damaged by the phytotoxicity of roridin A, whereas the *B. megapotamica* appears to be completely resistant to damage.

Another probable plant metabolite of a trichothecene has been recently reported as 4-O-cinnamoyltrichothecolone, produced in

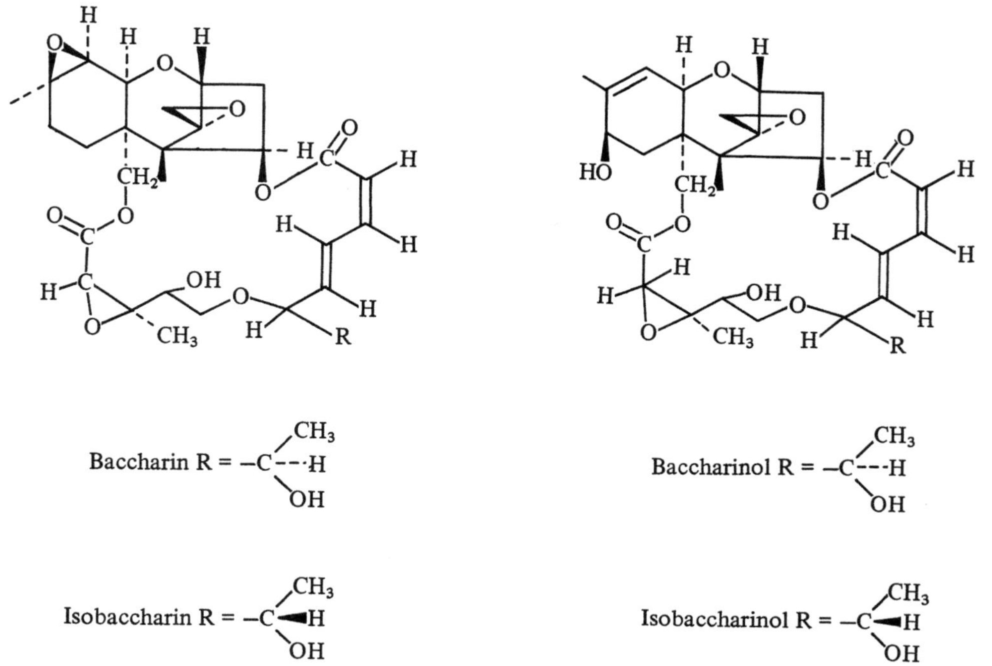

Fig. 8. The structure of the baccharinoids, transformation products of roridins by *Baccharis megapotamica* (Kupchan et al. 1977)

anise fruits naturally infected with *Trichothecium roseum* (Ghosal et al. 1982). Extracts of pure cultures of the fungus contained trichothecin, trichothecolone and 4-acetyltrichothecolon but the 4-O-cinnamoyltrichothecolone was only produced when the fungus was grown on anise fruits. Since anise fruits contain relatively high amounts of cinnamic acid and cinnamic acid esters, it is probable that host parasite interaction gives rise to the 4-O-cinnamoyltrichothecolone.

Many of the trichothecenes produced in culture are simple acyl derivatives of a parent polyol. Usually more than one of these compounds are found in the fungal culture extract, and many recent studies have attempted to elucidate the relationships and biological transformations which occur among the trichothecenes. Kotsonis and Ellison (1975) studied the production of T-2 toxin and HT-2 toxin (a 4-desacetyl derivative) in cultures of *Fusarium poae* grown at room temperature. T-2 toxin was produced prior to HT-2 and appears to be the precursor of HT-2 toxin in these cultures. Yoshizawa et al. (1980a) have done a more extensive study on T-2 toxin transformations by mycelia of *Fusarium graminearum*, *F. nivale*, *Calonectria nivale* and *F. sporotrichioides*. T-2 toxin was converted into acetyl T-2 by all strains except a T-2 toxin producing strain of *F. sporotrichioides* which converted T-2 toxin to HT-2 toxin and neosolaniol. In a similar fashion, HT-2 toxin was acetylated at the 3 position by *F. graminearum* and *C. nivale;* however, *F. nivale* failed to produce 3-acetyl HT-2 toxin from HT-2

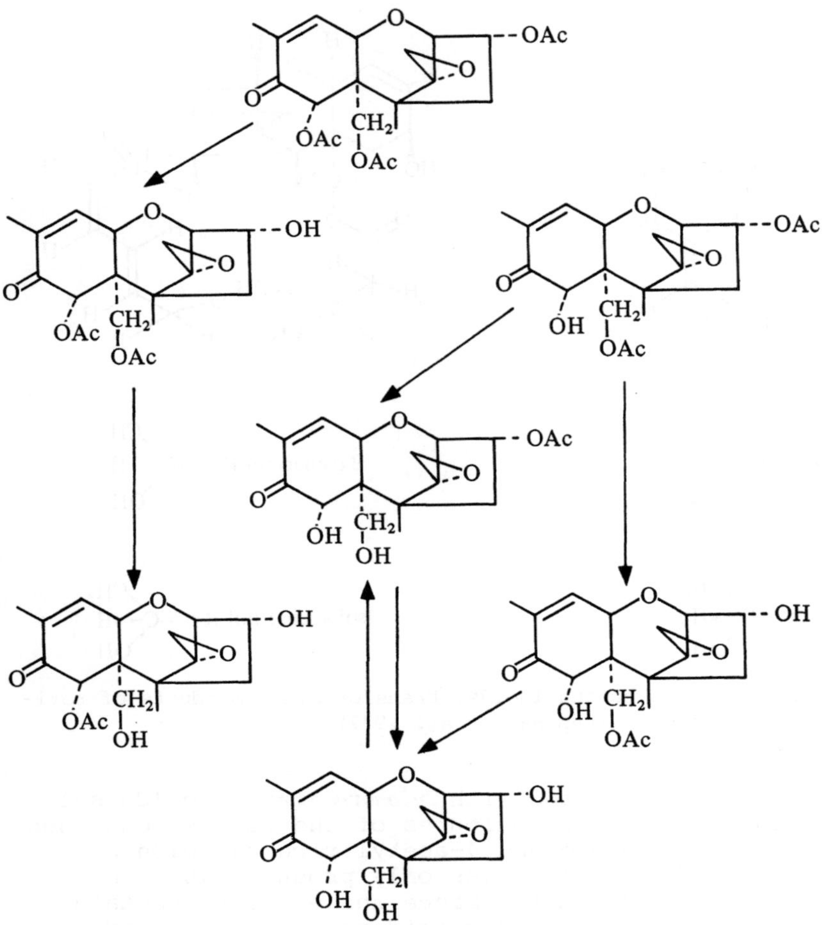

Fig. 9. Transformations of the triacetate and diacetate of deoxynivalenol catalyzed by *Fusarium* spp. Data are from Yoshizawa and Morooka 1977

toxin. *F. nivale* and *F. graminearum* would also convert 3-acetyl HT-2 toxin into HT-2 toxin. From results of these studies it appears that microbial acetyl conjugation of trichothecenes occurs predominantly at the C-3α position. Likewise Yoshizawa and Morooka (1975 and 1977) demonstrated that *F. roseum*, *F. nivale* and *F. solani* converted 3-acetyldeoxynivalenol into deoxynivalenol. Deoxynivalenol triacetate was also deacylated by *F. solani* into 7, 15-diacetyldeoxynivalenol which was further deacylated to give 7-acetyldeoxynivalenol. Starting with deoxynivalenol, only a small amount of acetylation which led to the formation of 3-acetyldeoxynivalenol could be demonstrated by intact mycelium of *F. nivale*. These fungal transformations of deoxynivalenol are summarized in Fig. 9.

The type of transformations shown in Fig. 9 readily explain the mixtures of trichothecenes which are usually found together in pure fungal cultures (Ueno et al. 1973a). Thus, fungal isolates which produce class A trichothecenes might produce neosolaniol, T-2 toxin, HT-2 toxin and diacetoxyscirpenol, whereas those isolates which synthesize 8-keto derivatives (class B) might produce fusarenon-x, nivalenol, diacetylnivalenol, etc.

Claridge and Schmitz (1978) studied the microbial transformation of diacetoxyscirpenol by *Streptomyces griseus*, *Mucor mucedo*, and *Acinetobacter calcoaceticus*. Deacetylation at the 4 position was brought about by *A. calcoaceticus*, while deacetylation at C15 occurred in *S. griseus*. *M. mucedo* carried out acetylation in the 3 position starting with either diacetyoxyscirpenol, 4-acetoxyscirpendiol, and 15-acetoxyscirpendiol. Cultures of *F. oxysponum* would also carry out acetylation in the 3 position of diacetoxyscirpenol, but the major product was 3-acetoxyscirpene-4,-15-diol (Claridge and Schmitz 1979).

All of the above transformations appear to be enzymatically catalyzed since control experiments were done to rule out uncatalyzed hydrolysis. Each organism capable of achieving these transformations seems to have specific acylating enzymes. It thus appears that many of the known trichothecenes arise in fungal culture due to these types of transformations. It is not surprising, therefore, that pH and temperature can have such profound effects not only in the amount of trichothecenes produced in culture but also on the quantitative distribution of the toxins produced as well.

C. Detection of Trichothecenes

Development of methods for the detection and quantitative analysis of trichothecenes in fungal cultures and in naturally contaminated foodstuffs has been actively pursued since it was recognized that these compounds were of potential importance in human and animal disease (Bamburg et al. 1969). Many of the methods available and their advantages and disadvantages have been reviewed periodically during the past decade (Bamburg and Strong 1971; Bamburg 1972; Eppley 1975; Pathre and Mirocha 1977; Eppley 1979; Ueno 1980a). Several advances in physico-chemical and immunochemical methods for trichothecene analysis have been reported during the past 3 years. These improvements along with some new bioassays will be briefly discussed below.

I. Physical Methods of Analysis

While many analytical methods work well for quantitating trichothecenes in pure samples, most analytical methods must be applied to rather crude toxic extracts obtained from fungal cultures of foodstuffs. Since large quantities of contaminating oily residue are extracted along with the trichothecenes, many of the recent papers deal with procedures for cleaning up the sample prior to quantitative analysis. Stahr et al. (1979) presented a method for determining T-2 toxin, diacetoxyscirpenol and deoxynivalenol in foodstuffs which involved an acetonitrile-H_2O extraction, petroleum ether defatting, ferric chloride decolorization, chloroform partitioning, and concentration. Gas chromatography (GC) and thin-layer chromatography (TLC) were used to separate the major trichothecenes. The gas chromatographic procedure is sensitive to 10 ng while the TLC method requires a minimum of 100 ng of toxin. Maximum sensitivity based on weight of starting material is 1 ppm (1mg/kg). Kamimura et al. (1981) also reported a method for simultaneous detection of several trichothecenes in cereals, grains and other foodstuffs. Following methanol extraction, a two step chromatographic method was used to partially purify the trichothecenes. Each mycotoxin in the mixture was quantitated by TLC and GC. The minimal detectable levels in the foodstuffs were 2 µg/kg (2 ppb) for nivalenol, deoxynivalenol, and fusarenon-x, and 80 µg/kg diacetoxyscirpenol, neosolaniol, T-2 toxin and HT-2 toxin. Recoveries of the toxins averaged 85%. The improved sensitivity on GC detection arose through the use of an electron capture detector which was 3 to 50x more sensitive for the derivatized trichothecenes than the flame ionization detector.

By far the best method for trichothecene analysis in foodstuffs appears to be GC-mass spectrometric analysis. Mirocha and coworkers (1976) have applied GC-mass spectral analysis to trichothecenes and have used selective ion monitoring (SIM) to detect levels of T-2 toxin and deoxynivalenol at levels of less than 50 ppb in naturally contaminated feed. Parales et al. (1975) have used the GC-SIM technique for quantitation of T-2 toxin in milk. A complete mass spectrum of the suspected trichothecene can be obtained for conformational purposes. Szathmary et al. (1980) developed a glass capillary GC column which gives very high resolution of the four trichothecenes analyzed on it. Coupling this column separation with mass spectral analysis to confirm the identification of the component in the sample peak allows one to bypass many of the sample clean-up procedures. Sensitivity of the method is of the order of 50 µg/kg or 50 ppb of toxin in feed. Scott et al. (1981) reported even lower detection limits (10 ppb) for deoxynivalenol using a GC-MS system but applying a significant clean-up procedure to the sample extract before analysis. Similar results (detection limit of 5 ppb) for T-2 toxin in corn extracts has been reported by Chaytor and Saxby (1982).

A less sensitive but useful separatory tool for trichothecenes has been reverse phase HPLC. A detection limit of about 1 µg per injection for T-2 toxin was obtained using a differential refractometer as a detector (Schmidt et al. 1981). Amounts of toxin up to 1 mg can be separated in this way. T-2 toxin, HT-2 toxin and diacetoxyscirpenol have all been separated by this method and, following a preliminary sample work-up, T-2 toxin and HT-2 toxin have been identified in extracts of moldy rice. The major advantages of HPLC over gas chromatography is that no sample derivatization is necessary.

A more useful application of HPCL involves combining the separation capabilities of this method with the identification capabilities of field desorption and electron impact mass spectrometry. Schmidt et al. (1982) have used this technique to identify acetyl-T-2 toxin in extracts of moldy rice.

Thin-layer chromatography (TLC) has always been a favorite method for mycotoxin analysis because of its simplicity and low cost. Trichothecene analysis by TLC has suffered because of the lack of specific detection methods that were sensitive in the nanogram range. Sano et al. (1982) recently described a new fluorodensitometric method for quantitating trichothecene mycotoxins in the range of 10-1500 ng per spot. The detection method is sensitive for epoxy compounds such as the trichothecenes and involves reaction first with nicotinamide at elevated temperatures, then with 2-acetylpyridine followed by potassium hydroxide solution, and finally with formic acid solution. Trichothecenes appear as light blue fluorescent spots under UV light (360 nm). Unfortunately, the presence of other alkylating agents in extracts of natural foodstuffs still requires that some sample clean-up procedures be used.

The 8-keto trichothecenes show electrochemical activity in the polarograph and Palmisano et al. (1981) have developed a differential-pulse polarography technique sensitive enough to detect as little as 50 ng/g of deoxynivalenol in extracts of corn. The detection limit for a pure 8-keto trichothecene such as deoxynivalenol was 8.6 ng/ml, but like many other procedures the extracts of naturally infected foodstuffs required substantial clean-up before analysis.

II. Immunochemical Methods of Analysis

Chu et al. (1979) first reported the production of an antibody directed against a trichothecene, T-2 toxin. The antibody was prepared in rabbits immunized with a bovine serum albumin-T-2 toxin hemisuccinate conjugate. The antibody was quite specific and sensitive for T-2 toxin. Using a level of [^{3}H]-T-2 toxin which gave 50% of the saturating level of binding, Chu et al. (1979) found that T-2 toxin gave 50% inhibition of this binding at 3.5 ng/assay. To achieve the same level of inhibition, 5.7 times as much HT-2 toxin and 37 times as much T-2 triol were required. All other trichothecenes tested (including neosolaniol, T-2 tetraol and 8-acetylneosolaniol) were required in at least 400 fold greater amounts than T-2 toxin to achieve 50% inhibition of the [^{3}H]-T-2 toxin binding. This T-2 toxin antiserum has been used in a radioimmunoassay for quantitation of T-2 toxin in urine and milk (Lee and Chu 1981a) as well as in wheat and corn (Lee and Chu 1981b). While some preliminary sample clean-up was needed for all of these assays, the clean-up procedures were far less extensive than those required for some of the less specific physical methods of analysis. Recoveries of the toxin in spiked samples was 80% or better and detection limits in the range of 0.5 to 2.5 ppb were obtained.

The availability of antibodies specific for T-2 toxin also led to the development of a sensitive enzyme-linked immunosorbant assay (ELISA) (Pestka et al. 1981). In this type of assay, T-2 toxin antiserum is dried onto polystyrene microtissue culture plates and the ELISA conducted by simultaneously incubating samples containing T-2 toxin (or unknowns) with a T-2 toxin-horseradish peroxidase conjugate. Competition curves could then be prepared for the total amount of bound enzyme. This assay appears to be almost as sensitive as the radioimmunoassay (minimal levels of about 2.5 pg T-2 toxin/assay) but has the advantages of not requiring radioisotopes or expensive scintillation counters. However, all of these immunochemical assays are useful in detecting only one specific trichothecene and not useful tools for screening for trichothecenes in general. Perhaps utilization of a less derivatized trichothecene antigen such as trichodermol might give antibodies which are more specific in recognizing the trichothecene ring system rather than a particular derivative.

III. Biological Methods of Analysis

Of the various biological assays for trichothecenes, none has surpassed the skin test for simplicity and reliability (Wei et al. 1972). The detection limit on the skin of rabbits is about 10 ng/test for the most dermally active trichothecenes (Chung et al. 1974), but all trichothecenes which have been examined give some degree of response in this assay (Ueno 1980a). The quantitative nature of the assay can be improved by directly assessing the intensity of a reaction from an unknown solution to a graded series of standards applied to the same animal (Hayes and Schiefer 1979).

New biological tests for trichothecenes based on toxicity to specific organisms have appeared, but none are specific for trichothecenes and all may give false positive tests with extracts from naturally infected foodstuffs (Prior 1979; Siriwardana and Lafont 1978). A whole animal assay for trichothecenes based on the rejection or acceptance of spiked drinking water was developed for mice (Burmeister et al. 1980). Several mycotoxins were utilized in this assay but only trichothecenes caused water refusal. Levels of 2 μg/ml were required to get a significant response.

D. Trichothecene Production

I. Distribution of Trichothecene Producing Fungi

The majority of trichothecene producing fungal isolates belong to the genus *Fusarium*. Of the nine species of *Fusaria* recognized by Snyder and Hansen (1945), eight have been reported to produce trichothecenes. *Fusarium* nomenclature is especially confusing due to the seven different classification systems proposed during the past 60 years (Wollenweber and Reinking 1935; Snyder and Hansen 1945; Tsunoda et al. 1968; Bilai 1970; Booth 1971; Joffe and Palti 1975). Ueno (1980a) has recently reviewed the taxonomy of *Fusarium* and has attempted to clarify synonomous species between some of the major classification systems.

In addition to the *Fusaria*, trichothecenes have been isolated from species of *Myrothecium*, *Trichothecium*, *Trichoderma*, *Stachybotrys*, *Calonectria* (perfect stage of *F. nivale* and *F. rigidiusculum*), *Cephalosporium*, *Cylindrocarpon*, and *Verticimonosporium*. Most of these fungi are common air-borne and soil organisms and appear to be ubiquitous in nature. Trichothecene producing strains of these fungi have been isolated from moldy grains, fruits, and soil found in Africa (Böhner et al. 1965), Austria (Vesonder and Ciegler 1979), Canada (Greenway and Puls 1976; Vesonder and Ciegler 1979; Scott et al. 1980; Andrews et al. 1981; Davis et al. 1982), Czechoslovakia (Betina and Vankova 1977), Finland (Ilus et al. 1977), France (Bamburg et al. 1968a; Jemmali et al. 1978), Germany (Marasas et al. 1979a), Hungary (Gyimesi and Melera 1967; Harrach et al. 1981), India (Ghosal et al. 1976; Rukmini and Bhat 1978; Ghosal et al. 1978), Italy (Bottalico 1977), Japan (Tatsuno 1968; Ueno et al. 1969), New Guinea (Godtfredsen and Vangedal 1965), Scotland (Petrie et al. 1977), South Africa (Steyn et al. 1978; Marasas et al. 1979b), Soviet Union (Olifson 1957; Joffe 1962; Mirocha and Pathre 1973), United States (Bamburg et al. 1968a,b; Hsu et al. 1972; Mirocha et al. 1976; Vesonder et al. 1979a), and Thailand (Okuchi et al. 1968).

Although as a general rule fungi that produce the class B trichothecenes (8-keto) do not produce class A compounds and vice versa (Ueno et al. 1973a), exceptions have been reported. Ueno (1980a) reported an isolate of *F. lateritium* which produced both diacetoxyscirpenol and diacetylnivalenol. In addition, *F. roseum* isolates which can produce deoxynivalenol or various scripenol derivatives have also been found. In naturally infected grain it is not unusual to find many different trichothecene-producing fungi growing together, and thus one might expect to find class A and B toxins occurring together. Indeed, nivalenol, deoxynivalenol and T-2 toxin have been identified in extracts prepared from corn samples obtained in France (Jemmali et al. 1978). However, this finding appears to be more of an exception than the rule. Trenholm et al. (1983) looked for deoxynivalenol, diacetoxyscirpenol, neosolaniol, T-2 toxin, and HT-2 toxin in samples of Canadian white winter wheat. Of 41 samples tested, 16 contained detectable levels of deoxynivalenol but no samples contained any of the other trichothecenes. Therefore, the finding of these same toxins together in a single sample of "yellow rain" from a leaf might be considered unusual for natural contamination, though what might be natural for southeast asia has yet to be defined.

Trichothecium roseum is a common parasitic fungus of fruit and *T. roseum* has been isolated from a variety of moldy fruits including anise (Ghosal et al. 1982) and apple (Betina and Vankova 1977) Trichothecin, which had been produced in the apple by the fungus, was identified as the component responsible for the bitter taste.

II. Trichothecene Production in Culture

Improved yields of T-2 toxin can be obtained in shorter times by culturing the *F. tricinctum* on nutrient broth-moistened vermiculite incubated at 19°C (Cullen and Smalley 1981). A single isolate of *F. tricinctum* (NRRL 3299) yielded 339 mg T-2 toxin/l cultured in this way compared to 2.7 mg T-2 toxing/g when cultured on corn, 1.7 mg/g cultured on rice, 15.5 mg/l cultured on Gregory's medium, or 76 mg/l cultured on peptone supplemented Czapeks-Dox. This same isolate of *F. tricinctum* was shown previously by Burmeister (1971) to produce higher yields of T-2 toxin when grown on white corn grits. After culturing 3 weeks at 15°C on 1.2 kg of grits, at least 9 g T-2 toxin was produced. A lesser amount was produced on rice and none was produced on wheat indicating the substrate specificity for trichothecene production. Mass production of T-2 toxin and neosolaniol in shake cultures and jar fermentors was reported by Ueno et al. (1975). *F. solani* M-1-1 cultured at 27°C for 5 days yielded 440 mg/l crude toxin from which 170 mg/l T-2 toxin and 50 mg/l of neosolaniol were crystallized. These relatively high yields of trichothecenes produced by fungi grown on inexpensive substrates indicates that relatively large amounts of the toxin could be produced at very little expense. A single silica gel chromatography step is all that is required for separation of the trichothecenes. The pure toxins can be crystallized from the column fractions (Ueno et al. 1975).

An alteration in the growth conditions for the trichothecene-producing fungus can alter the nature of the mycotoxins produced. Bamburg and Strong (1969) reported that a particular isolate of *F. tricinctum* grown at 8°C produced mainly T-2 toxin whereas the same organism grown at 25°C produced mainly HT-2 toxin. Maximum production of Fusarenon-X by *F. nivale* has been reported to occur at 27°C in 6-8 days (Ueno et al. 1970a). The yield decreased at lower temperatures. Vesonder et al. (1982) studied deoxynivalenol-producing isolates of *Fusarium*. Maximum production of deoxynivalenol by *F. graminearum* on corn occurred at 30°C in 40 days, whereas an *F. roseum* isolate produced maximal amounts of deoxynivalenol at 26°C in 41 days. Less than 10% of the maximum amounts were produced when either culture was incubated at 15°C for up to 60 days.

III. Trichothecenes in Naturally Infected Foodstuffs

Outbreaks of animal diseases associated with mycotoxin production in animal feed occurs in episodes in many countries with the major problems usually occurring after a cold and wet harvest season. The problem arises when the moisture content of grains used for animal feed has not decreased sufficiently to prevent mold growth. Hsu et al. (1972) were the first to actually quantitate the levels of a trichothecene (T-2 toxin) in feed associated with a natural outbreak of moldy corn toxicosis. They identified 2 ppm of T-2 toxin in dried corn by a combination of gas chromatography and mass spectrometry. Since that time, diacetoxyscirpenol, deoxynivalenol (vomitoxin) and nivalenol (in addition to T-2 toxin) have been identified in field samples of grains as naturally occurring trichothecenes (Table 1).

Trichothecenes have also been detected in fruits such as apple (Betina and Vankova 1977) and anis (Ghosal et al. 1982), but only from this latter source have they been isolated and posi-

Table 1. Natural occurrence of trichothecenes by location

Country	Trichothecene	Amount(μg/g)	Substrate	Reference[a]
Austria	Deoxynivalenol	1.3-7.9	Corn	[1]
Brazil	Verrucarins/roridins	Unknown	Soil[b]	[2]
Canada	T-2 toxin	25	Barley	[3]
	Deoxynivalenol	7.9	Corn	[1]
	Deoxynivalenol	8.5	Wheat	[4]
France	T-2 toxin	0.02	Corn	[5]
	Nivalenol	4.28	Corn	[5]
	Deoxynivalenol	0.6	Corn	[5]
Germany	Diacetoxyscirpenol	31.5	Corn	[6]
Hungary	Diacetoxyscirpenol	1.5	Corn	[7]
India	Diacetoxyscirpenol	14	Corn	[8]
	T-2 toxin	4	Corn	[8]
	T-2 toxin	Unknown	Sorghum	[9]
	Trichothecin	62	Anise	[10]
	Trichothecolone	19	Anise	[10]
	4-acethyltrichothecolone	14	Anise	[10]
	4-cinnamoyltrichothecolone	35	Anise	[10]
Japan	Nivalenol	7.3	Barley	[11]
	Nivalenol	0.1-22.9	Barley	[12]
	Deoxynivalenol	5	Barley	[13]
South Africa	Deoxynivalenol	0.25-4	Corn	[14]
USA	T-2 toxin	2	Corn	[15]
	T-2 toxin	0.076	Mixed feed	[16]
	Diacetoxyscirpenol	0.5	Mixed feed	[16]
	T-2 toxin	1	-	[17]
	Diacetoxyscirpenol	10-50	-	[17]
	Deoxynivalenol	10-50	-	[17]
	Deoxynivalenol	8	Corn	[18]
	Deoxynivalenol	40	Corn	[19]
	Deoxynivalenol	0.5-10	Corn	[20]
	Deoxynivalenol	37	Corn	[21]
	Deoxynivalenol	0.1-1.8	Corn	[22]
	Diacetoxyscirpenol	0.38-0.5	Mixed feed	[22]
Zambia	Deoxynivalenol	7.4	Corn	[23]

[a] [1] Vesonder and Ciegler 1979; [2] Jarvis et al. 1981c; [3] Puls and Greenway 1976; [4] Trenholm et al. 1981; [5] Jemmali et al. 1978; [6] Siegfried 1977; [7] Szigeti 1976; [8] Ghosal et al. 1978; [9] Rukmini and Bhat 1978; [10] Ghosal et al. 1982; [11] Morooka et al. 1972; [12] Yoshizawa and Morooka 1973; [13] Yoshizawa and Morooka 1977; [14] Marasas et al. 1979b; [15] Hsu et al. 1972; [16] Mirocha et al. 1976; [17] Stahr et al. 1978; [18] Ishii et al. 1975; [19] Vesonder et al. 1976; [20] Vesonder et al. 1978; [21] Vesonder et al. 1979a; [22] Pathre and Mirocha 1979; [23] Marasas et al. 1978

[b] By inference

tively identified (Table 1). In addition, the finding of metabolic transformation products of the verrucarins and roridins in the leaf of the Brazilian shrub *Baccharis megapotamica* (Kupchan et al 1976, 1977) indicates that these compounds must have occurred naturally in the soil (Jarvis et al. 1981c).

As a result of sporadic outbreaks of mycotoxicoses, mycotoxin screening programs have been set up in both the United States and Canada. While a large number of fungal isolates obtained from hay (Davis et al. 1982) and cereal grains (Scott et al. 1980) produced T-2 toxin when cultured on autoclaved corn or on sterile rye grain, actual field production was confined to a restricted region of the country. The Canadian surveys have identified deoxynivalenol (0.14-8.5 ppm) in many samples of winter wheat in the Ontario region while T-2 toxins (up to 25 ppm) were found in 21 out of 200 samples tested in British Columbia. The levels detected were high enough to be contributing factors in feed refusal and in some of the ill effects (Andrews et al. 1981; Trenholm et al. 1981).

From the standpoint of human and animal disease, of equal interest to the occurrence of trichothecenes in feed is what happens to the toxins in the processing of foodstuffs. A recent study on this topic was carried out by Collins and Rosen (1981) who looked at the distribution of T-2 toxin in wet-milled corn products. Three concentrations of T-2 toxin ranging from 0.5 ppm to 8.7 ppm were used. About 2/3 of the toxin was removed by the steep and process water, 4% was in the starch, and the remainder was about equally distributed between the germ, gluten and fiber. The extraction process tended to concentrate the toxin in the germ which contained 163-194% of the toxin concentration found in the starting whole corn. Thus food products destined for human consumption made from the germ have the greatest chance of causing human health problems. Animal feed supplemented with concentrates of the corn steep liquor also could cause problems. T-2 toxin appears to be destroyed in the acidic or alkaline conditions used for preparing corn syrup and corn oil, respectively, and, therefore, along with the corn starch which contains little of the initial T-2 toxin, these corn products probably present less of a health hazard than the germ.

E. Human and Animal Mycotoxicoses in which Trichothecenes are the Suspected Causative Agents

I. Mycotoxicoses of Historical Interest

Diseases of fungal origin have plagued man and his domesticated animals for hundreds of years. Organized research into the identification of the causative fungi and attempts at elucidating the chemical principles involved were first undertaken by the Soviet Union in the 1930's and by Japan shortly thereafter. Interest in mycotoxins in the western world lagged considerably, and it was not until the discovery that the compound (aflatoxin) in peanuts which killed thousands of turkeys in Great Britain ("Turkey X disease") was of fungal origin and that it could cause cancer in experimental animals, that much attention was directed toward the role of mycotoxins in disease. While it is not possible to go back and analyze samples which caused severe myotoxicoses in previous decades using today's technology, comparative studies of the fungi and symptoms of the unfortunate victims of these diseases allow us to infer with some feeling of assurance that many of the unidentified mycotoxicoses described in the earlier literature arose through the ingestion of trichothecenes. One such disease, described by Woronin in 1891, is called "staggering grain toxicosis" and occurred in eastern Siberia. Symptoms of nausea and vomiting, along with visual disturbances, were common in humans who consumed the toxic millet or barley, and farm animals refused to eat the infected grain. The fungi isolated included *F. roseum*, a known trichothecene producing organism. The symptoms of vomiting and feed refusal are seen today in animals fed grain contaminated with either deoxynivalenol (also known as vomitoxin) which is produced by some species of *F. roseum*, or T-2 toxin, produced by several different *Fusarium spp.*

A second fungal disease, first described in 1937 in the Soviet Union is called dendrochiotoxicosis. This disease was originally a disease of horses but later was also described as affecting humans who worked in cotton factories. Lesions of the skin, especially around the mouth and nose were prevalent in both horses and humans while hemorrhagic symptoms were also reported in horses (Bilai and Pidoplisko 1970). The causative fungus was identified as *Dendrodochium toxicum*, and several toxic compounds were extracted from the fungus grown in pure culture. While structures for these compounds were never elucidated, their effects appear to be similar to the macrocyclic trichothecenes produced by *Myrothecium spp.* Tullock (1972) reported that *D. toxicum* is synonymous with *Myrothecium roridum;* thus, the isolated compounds probably represent members of the roridin family.

Because of the nature of the fungi isolated, trichothecenes have been implicated in Bean-hull poisoning of horses in Japan (Ueno et al. 1972a) as well as in a similar disease called leukoencephalomalacia which also affects members of the equine family (Pienaar et al. 1981). The major symptoms of this disease are the liquifactive necrotic lesions in the white matter of the cerebral hemispheres. While trichothecenes may certainly be involved in some aspects of this disease, the major causative organism has been identified as *F. verticillioides* (Sacc) Nirenberg a non trichothecene producing *Fusarium*. This fungus produces moniliformin, another mycotoxin which may be responsible for the disease.

Evidence is more direct for the involvement of trichothecenes in some other human and animal mycotoxicoses and so these will be discussed individually below. It should be pointed out, however, that most mycotoxicoses are very complex diseases and that seldom is only a single mycotoxin involved. In addition, the symptoms of the disease may be different in laboratory animals than in the human or domesticated livestock, making it difficult to identify the causative agents. Finally, the total nutritional state of the animal receiving the toxin may play a large role in the symptoms which develop (McLean and McLean 1969). Since moldy food has lost much of its nutritional value (Marasas 1969; Schoental 1980a Hayes and Schiefer 1980), the toxicity of the feed and the symptoms of the animals receiving the feed may both be exacerbated.

II. Alimentary Toxic Aleukia

Alimentary toxic aleukia (ATA) was a widespread human disease recorded in the Soviet Union from the early 19th century but first recognized as a significant problem in the 1930's. Extensive accounts of this disease which killed hundreds of thousands of people in the Orenburg district of the Soviet Union during World War II are available (Forgacs and Carll 1962; Mayer 1953; Joffe 1962). The disease was caused by human consumption of moldy grain which overwintered in the field. The fungi primarily responsible were identified as *Fusarium sporotrichioides*, *F. poae*, *F. tricinctum* and *Cladosporium epiphylum*. The physiological manifestations of the disease as described by Joffe (1965) were skin lesions, leukopenia, agranulocytosis, necrotic angina, hemorrhagic diathesis, sepsis and exhaustion of bone marrow. The toxic factor responsible was extracted in organic solvents and had very high activity in the rabbit skin test (Olifson 1957). One steroidal compound called sporofusariogenin was isolated (Olifson 1960) and reported to be one of the major causes of the disease. Based on the fungi responsbile for the disease and the symptoms of ATA, Bamburg et al. (1969) postulated that ATA was probably caused by a trichothecene such as T-2 toxin. Ueno et al. (1972b) analyzed extracts of fungi which produced steroidal compounds and showed that these fungi were capable of synthesizing T-2 toxin. Mirocha and Pathre (1973) identified T-2 toxin in an actual sample of the toxic extract obtained from V.I. Balai in the Soviet Union. Yagen and Joffe (1976), working with isolates of *F. sporotrichiodes* and *F. poae* obtained in the Soviet Union, also identified T-2 toxin as the major skin irritant in culture filtrates. More recently these workers have isolated six steroidal compounds from these same cultures (Yagen et al. 1980), none of which was active in the skin test and all of which were relatively nontoxic. Administration of pure T-2 toxin to cats produces all signs, symptoms and pathological changes associated with ATA (Lutsky et al. 1978). Thus, it appears that T-2 toxin was primarily responsible for the skin necrotizing activity of the extracts and most probably for ATA itself.

III. Red-Mold Disease

The involvement of *F. roseum* and *F. nivale* in episodic outbreaks of "red-mold" disease in Japan and the similarity in symptoms to the "moldy corn toxicosis" seen in the midwestern United States prompted Bamburg et al. (1969) to postulate a role for trichothecenes in this disease. The following year, fusarenon-X and nivalenol, the major toxic metabolites produced by these fungi, were characterized as belonging to the trichothecene group (Tatsuno et al. 1969; Ueno et al. 1969; Grove 1970a, 1970b). The symptoms of this disease are vomiting, diarrhea, hemorrhage in the intestine, skin inflammation, feed refusal, and infertility (Ueno et al. 1971a). Like moldy corn toxicosis (see below), red-mold disease is common when cold and wet weather develops during the harvest season.

IV. Stachybotryotoxicosis

The early literature about Stachybotryotoxicosis was reviewed by Forgacs and Carll (1962) and Forgacs (1965). The disease was reported in the Ukraine of the Soviet Union in the early 1930's where thousands of horses died after consuming moldy straw. Forgacs et al. (1958) isolated toxic strains of the causative fungus *Stachybotrys atra* and demonstrated that the symptoms, which included anemia, hemorrhagic septicemia, irritation of the mouth, nose and throat, leukocytopenia and hemorrhagic diarrhea, could be brought on by feeding animals infected straw. Recently an outbreak of the disease was reported in South Africa where 109 out of 568 sheep died after consuming wheat, barley and rye straw contaminated with *Stachybotrys chartarum* (Schneider et al. 1979). *S. chartarum* is synonymous with *S. atra* (Korpinen 1974). The main clinical signs occurred in two states: an elevated body temperature, listlessness, epistaxis and intermittent hemorrhagic diarrhea during the first phase and a worsening anemia, leukopenia and less severe hemorrhaging with a terminally elevated temperature in the second phase. Atrophy and necrosis of the lymphoid tissues was one of the most salient histological findings. *Stachybotrys atra* causing toxic symptoms and death of cattle in the Indian province of Tamil Nodu was reported in 1975 (Rajendran et al. 1975) indicating that toxigenic strains of the fungus may be found widely distributed not only in Europe (Hintikka 1977) but in the cooler regions of Southern Africa and Asia as well. The disease also affected farm laborers when they were exposed to aerosols from handling intoxicated straw (Drobotko et al. 1945). The causative agents for stachybotryotoxicosis were hypothesized to be trichothecenes by Bamburg and Strong (1971) and this hypothesis has been proven correct by Eppley and Bailey (1973) who isolated several class C trichothecenes (macrocyclic esters) from cultures of *S. atra*. Five of these compounds have been separated and given the names Satratoxins C, D, F, G, and H. Satratoxin D was shown to be identical to roridin E (Eppley and Bailey 1973). The other Satratoxin structures which have been recently elucidated are given in Fig. 4.

V. Feed Refusal and Vomiting Factors in Feed

Periodic occurrences of vomiting and feed refusal in man and animals fed moldy grain have been reported for over half a century (Dounin 1926; Mains et al. 1930). These incidents are sporadic in nature but usually correspond to cold temperatures and excessive moisture during the harvest season (Ueno et al. 1974). Fungi of the genus *Fusarium* have been the most prevalent on this moldy grain (usually corn), with isolates of *F. roseum* being quite common (Kotsonis et al. 1975a). Ueno et al. (1971b) demonstrated that Fusarenon-X induced vomiting in ducklings, cats, and dogs. Vesonder et al. (1973) isolated an emetic principle which they called vomitoxin (4-deoxynivalenol) from *Fusarium* infected corn. However, the 8-keto trichothecenes (type B) are not as potent as the type A trichothecenes in inducing the vomiting response making the name vomitoxin somewhat of a misnomer (Ueno et al. 1974). These results were extended by Vesonder et al. (1979b) who found that the type A trichothecenes such as T-2 toxin are also more potent (almost 2x) as feed refusal factors for both rats and swine (Vesonder et al. 1981a, 1981b).

VI. Moldy Corn Toxicosis

Moldy corn toxicosis, one of the most complex of the mycotoxicoses, is a recurring problem in midwestern United States. As with many of the natural outbreaks of mycotoxicosis where trichothecenes have been implicated as a causative agent, moldy corn toxicosis usually occurs following cold and damp harvest seasons when the corn has not had time to dry sufficiently before harvest. Symptoms of the toxicosis in pigs and cattle are diarrhea, reduced milk production in dairy cattle, and oftentimes hemorrhagic lesions in the liver, stomach, heart, lungs, bladder, kidney and intestine. It is the hemorrhaging that gives this disease its characteristic feature (hemorrhagic syndrome), and massive blood loss through the intestine and into the abdominal cavity is often the cause of death. Hsu et al. (1972) were the first to identify a trichothecene, T-2 toxin, in extracts of moldy corn which caused hemorrhagic syndrome and death in a herd of dairy cattle. Kosuri et al. (1970) produced a hemorrhagic syndrome in a cow by daily intramuscular injection of T-2 toxin for over 2 months, and Weaver et al. (1978b) have demonstrated hemorrhaging in swine given diacetoyscirpenol via intravenous injection. However, oral administration of either the toxins or a crude fungal extract containing the toxins failed to produce hemorrhagic lesions (Patterson et al. 1979). More recent reports indicate that while trichothecenes and T-2 toxin in particular may be present in moldy feed responsible for hemorrhagic syndrome (Petrie et al. 1977), these compounds alone can not be the causative agent (Patterson et al. 1979; Weaver et al. 1978d; Weaver et al. 1980; Weaver et al. 1981). While it seems likely that these reports rule out T-2 toxin and diacetoxyscirpenol as the *sole* causative agents of the hemorrhagic syndrome, those of other trichothecenes may well be involved. In addition, the nutritional state of the animal may well determine whether or not hemorrhaging will occur. Schoental (1980a) has discussed the possible synergism of vitamin deficiencies and the presence of mycotoxins in the etiology of disease. Moldy feed, deficient in one or more essential nutrient, may, when ingested with trichothecenes over a prolonged period, result in the hemorrhagic syndrome. Perhaps the experimental diets given to swine and cattle along with the T-2 toxin or diacetoxyscirpenol were too nutritionally balanced for the hemorrhagic syndrome to develop. Alternatively, synergistic action of the trichothecenes and other fungal metabolities might be required for the expression of the hemorrhagic syndrome.

VII. Fusariotoxicosis in Poultry

Wyatt et al. (1972a) described a disease syndrome observed in several commercial broiler flocks characterized by raised yellowish-white lesions in the oral cavity and lesions on the feet and shanks. Up to 10% of the birds in the infected flocks died and growth rates of the others were depressed. A flock of fancy pigeons fed visibly moldy feed developed similar lesions. Wyatt et al. (1972b) demonstrated that identical oral lesions in the chicken could be produced in three weeks by feeding control feed which had been spiked with T-2 toxin between 4 and 16 ppm. No lesions appeared on the feet or shanks implying that these lesions are caused by something other than oral consumption of T-2 toxin. Direct skin contact with the trichothecenes through scratching the moldy feed may be required to bring about the skin lesions.

Puls and Greenway (1976) described a somewhat similar disease in geese that consumed barley which was subsequently shown to contain T-2 toxin. The birds developed head tremors and died within one day of being force fed the contaminated grain. Severe necrosis of the proventriculus and gizzard was observed with especially severe necrosis of the mucous membranes and cellular invasion of the submucous area. Mice fed the same barley *ad libitum* developed a deterioration of the hair coat but no animals died. The actual consumption of the contaminated feed by the mice was not reported.

VIII. Idiopathic Abortion in Farm Animals

Spontaneous abortions in farm animals have been observed for years but since no cause was ever found, these abortions were referred to as idiopathic. In the midwestern United States idiopathic abortion in swine is a common problem (Weaver et al. 1978a) and is usually associated with occurrence of other feed problems brought on by cold and wet harvest seasons. One estrogenic mycotoxin produced by *Fusarium* species, called zearalenone, has been well characterized (Mirocha and Christensen 1974). Zearalenone is most certainly responsible for the vulvo-vaginitis of swine but it alone can not explain the abortion problem. However, trichothecenes are often found with zearalenone in moldy grain (Jemmali et al. 1978; Marasas et al. 1979b), and so Weaver et al. (1978a, 1978c) looked at the effects of T-2 toxin on porcine reproduction and abortion. One sow injected intravenously with a single dose of 0.21 mg/kg T-2 toxin aborted 48 h post injection. While these large doses of T-2 toxin administered intravenously are equivalent to what might be consumed orally over a 2-3 month period, the fact that abortions occurred without any evidence of gross or microscopic lesions in either the sow or the fetuses suggests that T-2 toxin and perhaps other trichothecenes may play a significant role in the swine abortion problem. The feed refusal problem, emesis, moldy corn toxicosis, abortion and vulvo-vaginitis problems in swine are certainly all interrelated. Corn innoculated with *Fusarium* was allowed to mold in the field before harvesting in order to simulate natural mycotoxin producing conditions (Young et al. 1981). This corn developed levels of 39 ppm zearalenone and 37.5 ppm deoxynivalenol during storage and before feeding to animals. This corn was then mixed with healthy corn to which various levels of zearalenone and/or deoxynivalenol were added. The net result of these studies indicated that while zearalenone probably was the major estrogenic substance and was responsible for the changes in histology of the reproductive tract, neither zearalenone nor deoxynivalenol was primarily responsible for reduced feed consumption and that other factors are most likely involved (Young et al. 1981).

IX. Yellow Rain

The term "yellow rain" describes a sticky yellow powder released from munitions used against the Hmong tribesmen in Laos and Kampuchea (Seagrave 1981). Symptoms of exposure to this chemical warfare agent have been described as occurring in rapid sequence – "dizziness, nausea, coughing of blood tinged material, choking, vomiting of massive amounts of blood, shock, and death in those directly exposed to the powder. For those on the periphery of the attacks or who eat or drink contaminated food or water, symptoms take longer to develop (days rather than minutes to hours) ..." (State Dept. Release of 9/14/81). Analysis of leaf and stem samples of foliage from a region of attack, as well as pond water and rock scrapings from other areas where attacks were reported showed the presence of trichothecenes (T-2 toxin, nivalenol, deoxynivalenol and diacetoxyscirpenol). Although these compounds were not found in control samples from the same region of the country, and they were present in higher concentrations than have been reported for trichothecenes in naturally infected feedstuffs, they still only constituted up to 150 ppm in the samples analyzed. The question remains as to what constitutes the remainder of the mixture called yellow rain.

X. Dietary Influences and Symptoms of Mycotoxicoses

Schoental (1979a, 1980a, 1980b) has investigated the public health significance of the trichothecene mycotoxins, especially the possible involvement of these mycotoxins in previously unsuspected diseases such as pellagra, fetal alcohol syndrome, cobalt-beer cardiomyopathy and tumors of the upper alimentary tract. A good case was certainly presented for the involvement of mycotoxins in the expression of symptoms associated with pellagra (Schoental 1980a).

In a recent report, Hayes and Schiefer (1980) demonstrated that suppression of erythropoiesis in mice fed a diet containing 20 ppm T-2 toxin was temporary but that the interval between the suppression and the regeneration of erythroid precursor cells was dependent on the level of protein in the diet. Thus, the nutritional state of the animal certainly can play a role in the symptoms of the mycotoxicosis. No studies have as yet been published which combine the likely vitamin deficient state of animals with mycotoxin administration, but this is one area in which research should be actively pursued. It may be that only after these relationships are understood will it be possible to duplicate the symptoms of natural mycotoxicoses in experimental animals in a controlled environment.

F. Biological Activity of the Trichothecenes

While the specific activity of many of the trichothecene compounds differs significantly in tests of biological activity, the trichothecenes have many biological properties in common. These common properties include the toxicity of the trichothecenes to eukaryotic organisms and the dermatidic or skin-necrotizing activity which, because of its sensitivity, has often been used in the screening of samples for the presence of trichothecenes (Bamburg and Strong, 1971).

I. Phytotoxicity

Many of the trichothecene producing fungi are plant pathogens and the fungi production of the trichothecenes may be responsible for some of the symptoms of parasitic invasion. The shoots and leaves of intact plants were noticeably affected by direct application of several trichothecenes. Different species of plants differ remarkably in their susceptibility to trichothecenes. At 1 µg/ml diacetoxyscirpenol resulted in scorched foliage and inhibition of stem elongation in the pea, but mustard, wheat, beetroot and carrots were unaffected by 10 times this concentration (Brian et al. 1961). Table 2 sumarizes some of the toxocity studies done on plant seedlings with trichothecenes.

Translocation of trichothecenes from the leaves to the roots does not appear to occur in pea plants since application of diacetoxyscirpenol to the leaf has no apparent effect on other parts of the plant (Brian et al. 1961). However translocation from the roots to the leaves appears to occur with T-2 toxin and trichodermol (Marasas et al. 1971; Cutler and LeFiles 1978). The most dramatic uptake and translocation of trichothecenes was discovered in the Brazilian shrub *Baccaris megapotamica* (Kupchan et al. 1976; Jarvis et al. 1981c). This shrub is completely resistant to the effects of roridin A and verrucarin A and yet it will take up these compounds through the root system, transform them to their 8β-hydroxy-derivative and concentrate them in the leaf. Tomato, pepper and artichoke plants also show uptake and translocation of the compounds but these plants are severely damaged by the trichothecenes.

II. Cytotoxicity

Several different trichothecenes have been tested for cytotoxic activity against cultured cell lines and indeed this class of compounds contains some of the most cytostatic agents known. Verrucarin A had an ID_{50} for growth of cultured mouse P-815

Table 2. Phytotoxicity of trichothecenes

Trichothecene	Plant	Dose	Tissue	Effect	Reference[a]
Diacetoxy-scirpenol	Pea seedling	1 µg/ml	Foliage	Scorching	[1]
	Pea seedling	1 µg/ml	Stem	Reduced elongation	[1]
	Pea seedling	10 µg/ml	Whole plant	Death	[1]
	Lettuce	10 µg/ml	Whole plant	Death	[1]
	Winter tares	10 µg/ml	Whole plant	Death	[1]
	Cress	10 µg/ml	Whole plant	Stunting	[1]
	Tomato	10 µg/ml	Whole plant	Stunting	[1]
	Mustard	10 µg/ml	Whole plant	No effect	[1]
	Beetroot	10 µg/ml	Whole plant	No effect	[1]
	Wheat	10 µg/ml	Whole plant	No effect	[1]
	Carrot	10 µg/ml	Whole plant	No effect	[1]
	Cress seedling	10 µg/ml	Roots	ID_{50}	[1]
4, 15-diacetyl-nivalenol	Pea seedling	10 µg/ml	Foliage	Scorching	[1]
	Pea seedling	10 µg/ml	Roots	ID_{50}	[1]
T-2 toxin	Pea seedling	1 µg/ml	Roots	15% reduction in wt. & length	[2]
	Pea seedling	10 µg/ml	Whole plant	Death	[2]
Trichothecin	Bean	50 µg/ml	Leaf	Spot necrosis	[3]
	Tobacco	200 µg/ml		No effect	[3]
Trichodermin	Wheat	29 µg/ml	Coleoptile	80% inhibition in elongation	[4]
	Tobacco	29 µg/ml	Whole plant	72% growth inhibition at 28d	[4]
	Corn	29 µg/ml	Whole plant	No effect	[4]
	Bean	29 µg/ml	Whole plant	No effect	[4]
8-acetylneo-solaniol	Wheat	0.42 µg/ml	Coleoptile	Inhibition of elongation	[5]
Roridin A or Verrucarin A	Tomato	500 µg/ml	Roots	Dead in 3 days	[6]
	Pepper	500 µg/ml	Roots	Dead in 3 days	[6]
	Artichok	500 µg/ml	Roots	Dead in 3 days	[6]

[a][1] Brian et al. 1961; [2] Marasas et al. 1971; [3] Bawden and Freeman 1952; [4] Cutler and LeFiles 1978; [5] Lansden et al. 1978; [6] Jarvis et al. 1981c

cells of 0.6 ng/ml (Härri et al. 1962). Dividing cells were much more affected by the compound than were non-dividing cells, and mitosis was completely disrupted (Rüsch and Stähelin 1965). The cytostatic and cytotoxic nature of trichodermin (Godtfredsen and Vangedal 1965) diacetoxyscirpenol, diacetylnivalenol and trichothecin (Grove and Mortimer 1969), fusarenon-X (Ueno et al. 1969), crotocin, crotocol and trichothecolone (Gláz et al. 1966), and T-2 toxin (Bamburg 1972) have been reported. Table 3 summarizes the inhibition obtained for these different trichothecenes.

Table 3. Comparative toxicity and protein synthesis inhibitory properties of several trichothecenes[a]

Name	Class	ID_{50} (μg/ml) Protein synthesis	Cytotoxicity[b] (ng/ml)	LD_{50}[e] (mg/kg)
Verrucarin A	C	0.01	1.0	0.5 (ip)
Roridin A	C	0.01	1.0[c]	1.0 (iv)
T-2 toxin	A	0.03	1.0	3.0 (ip)
Diacetoxyscirpenol	A	0.03	5.0	1.1 (ip)[f]
HT-2 toxin	A	0.03		9.0 (ip)
4, 15-diacetylnivalenol	B	0.10	30	9.0 (ip)
Trichothecin	B	0.15	75	300 (iv)
15-monoacetylnivalenol	B	0.20		3.4 (ip)
Fusarenon-X	B	0.25	100[d]	3.3 (ip)
Neosolaniol	A	0.25		14.5 (ip)
7-hydroxydiacetoxy scirpenol	A	0.40	30	5 (ip)
Crotocin	D	0.45	250	700 (ip)
Trichodermin	A	1.0	75	500 (sc)
4-deoxynivalenol	B	2.0		70 (ip)
Nivalenol	B	3.0	225	4.0 (ip)
3-acetyldeoxynivalenol	B	10.0		49 (ip)
Tetraacetylnivalenol	B	10.0	250	
Trichothecolone	B	20.0	5000	100 (iv)

[a]Data from: Grove and Mortimer 1969; Grove and Hosken 1975; Ueno 1977; Yoshizawa and Morooka 1977; Ohta et al. 1978; Härri et al. 1962
[b]Unless otherwise indicated these are the lowest toxic doses for HEp2 cells
[c]ID_{50} for mouse p815 cells
[d]Lowest toxic dose for HeLa cells
[e]In mice
[f]In rats

III. Insecticidal and Larvacidal Activity

Trichothecenes show toxicity to mosquito larvae (*Aedes aegypti*) (Grove and Hoskin 1975) and *Drosophilia melanogaster* larvae and adults (Reiss 1975). In general, the larvicidal activity toward mosquito eggs parallels the cytotoxicity of the trichothecene with the verrucarins A and B, roridin H, acetylverrucarin A, trichothecin, nivalenol, and T-2 toxin all causing greater than 50% mortality at doses of 25 μg/ml in 72 hrs (Grove and Hosken 1975).

IV. Structure-Activity Relationships

The structure-activity relationships among the naturally occurring and chemically modified trichothecenes were reviewed by Bamburg and Strong (1971) with regard to cytotoxicity and toxicity to rats. The presence of the 12,13-spiroepoxide was necessary for biological activity, reduction of the 9-10 double bond in the A ring decreased the toxicity by 75-80% from the parent compound. Derivatives in which the epoxide group remained but which had a cleaved ring C so as to permit nucleophilic attack on the rear side of the epoxide were also not cytotoxic (Grove and Mortimer 1969). Similar findings have been reported for the larvicidal activity of the trichothecenes (Grove and Hosken 1975). Both the larvicidal and cytotoxic activities of the trichothecenes show a dependence upon the degree and nature of the acyl substituents. In part, these substituents influence the toxicity of these compounds by improving their uptake into cells by increasing their lipid solubility. However, in most cases, the fully acylated compound is not the most toxic indicating that properties other than lipid solubility are important. These properties are probably the same as those which influence ribosome binding and inhibition of protein synthesis, a property of the trichothecenes which correlates fairly well with their toxicity (Table 3). These properties are described in detail in Section G.II.

G. Biochemical Studies on the Action of Trichothecenes

I. DNA Synthesis

The inhibition of DNA synthesis by trichothecene mycotoxins was first reported in Ehrlich ascites tumor cells (Ueno and Fukushima 1968) and HeLa cells (Ohtsubo et al. 1968). No inhibition of RNA synthesis was observed during short exposures of the cells to concentrations of nivalenol which inhibited DNA synthesis and protein synthesis almost completely. Other workers have reported partial inhibition (86%) of RNA synthesis in the presence of levels of trichodermin which inhibit protein synthesis by 97%, a pattern of inhibition typical of protein synthesis inhibitors (McLaughlin et al. 1977). Recent studies on the content of DNA and protein in cultured Chinese hamster ovary cells treated with diacetoxyscirpenol for 12 hours have shown that the protein/DNA ratio remains constant for cells in every phase of the cell cycle (Teodori et al. 1981). Thus protein synthesis and DNA synthesis seem to be affected to an equal extent. Since the trichothecenes demonstrated no inhibitory effects on DNA polymerase, thymidine kinase and thymidylate kinase in vitro, the inhibition of DNA synthesis probably arises from a secondary effect of the inhibition of protein synthesis (Ueno 1980a).

II. Protein Synthesis

The inhibition of protein synthesis in a rabbit reticulocyte lysate by the trichothecene nivalenol was first reported by Ueno et al. in 1968. Since then, a wide variety of trichothecenes has been tested for inhibition of protein synthesis on eukaryotic ribosomes (Ohtsubo et al. 1972; Ueno et al. 1973b; Carrasco et al. 1973; Stafford and McLaughlin 1973; Tate and Caskey 1973; Cundliffe et al. 1974; Barbacid and Vazquez 1974; Wei et al. 1974; Wei and McLaughlin 1974; Schindler 1974; Mizuno 1975). The mechanism of this inhibition and the structure-function relationships among the trichothecenes have been reviewed (Cundliffe and Davies 1977; McLaughlin et al. 1977; Carter and Cannon 1977, 1978). The trichothecenes can be divided into 3 classes based upon their ability to inhibit chain initiation (I), chain elongation (E) and/or chain termination (T). In addition, the chain initiation inhibitors can be further divided into two subclasses – those that only inhibit the function of intact ribosomes (I_2) and those that can also prevent formation of the 80S initiation complex (I_1) (Cundliffe and Davies 1977). Carter and Cannon (1978) showed that the expression of either E or I type behavior is dependent upon the concentration of some trichothecenes, thus allowing them to be further subdivided into classes of pure I, E and T type behavior as well as I type showing partial E type behavior at high concentration (I_2-E) and E-type showing I type behavior at low concentration (E-I_2).

Doyle and Bradner (1980) have reviewed the classification of the trichothecenes into different protein synthesis inhibitor groups and have summarized the structural features in common among the trichothecenes in each category. The termination inhibitors (trichodermol and some of its derivatives) are either unsubstituted or carry only a small substituent at the C4 position. Substitution by an ester group on the C4 position only, will make the compounds elongation inhibitors. However, some elongation inhibitors have no acyl group on the C4 position but have substituents elsewhere. An ester function on the C15 position will make a compound that showed pure E type behavior become an E-I_2 type compound. Compounds show I type behavior if, in addition to an ester at C15, either the C_3 or C_4 alcohol groups are acylated. The macrocyclic trichothecenes all belong in this category and, furthermore, are all I_1 type inhibitors. Table 4 shows the trichothecenes which have been categorized according to their behavior in the rabbit reticulocyte protein synthesis system and the structural features shared by each class.

Through studies on the mechanism of action of trichothecenes, Schindler et al. (1974) identified a mutant yeast which was resistant to the action of trichodermin. These workers identified the mutation as belonging to a component of the 60S ribosome. This mutant is in fact, resistant to the action of all the trichothecenes which indicates these compounds probably all bind at an identical site on the ribosome (McLaughlin et al. 1977). Grant et al. (1976) showed the reistance to trichodermin is conferred by a single gene, tcml, located on chromosome XV. Lo et al. (1980) showed that trichodermin resistant yeast have an altered ribosomal protein. Fried and Warner (1981) have cloned the gene responsible for trichodermin resistance and the gene specifies the ribosomal protein 1 (L3), a component of the 60S subunit. All eukaryotic ribosomes that have been examined have a large protein subunit homologous to the L3 protein in yeast (Fried and Warner 1981), and it seems likely that this protein will prove to be the common site for eukaryotic protein synthesis inhibition by the trichothecenes.

Table 4. Effects of trichothecenes on protein synthesis in eukaryotes

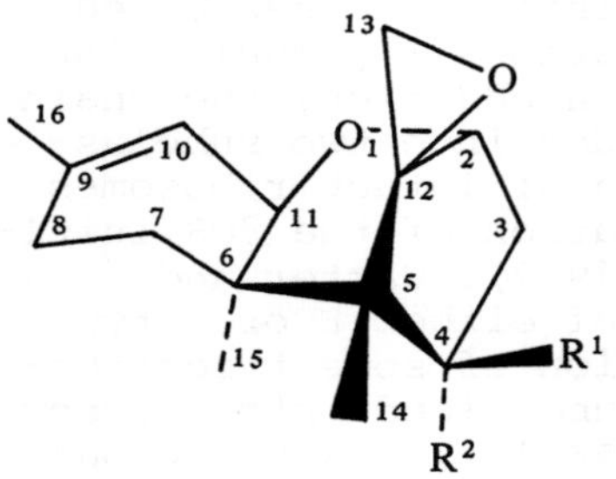

Termination inhibitors

	R_1	R_2	Class
Trichodermol	OH	H	T
Trichodermone	O	-	T

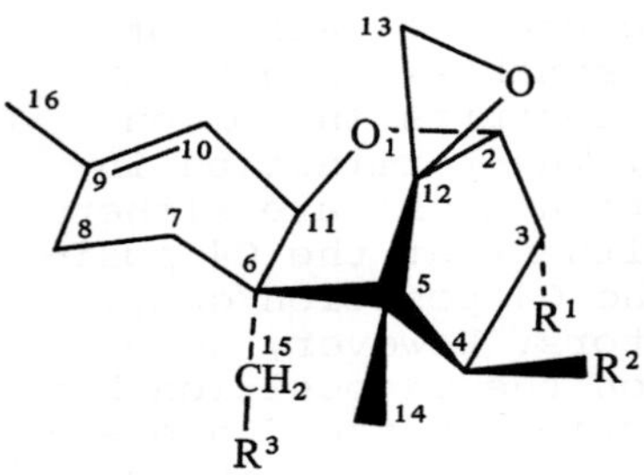

Elongation inhibitors

	R_1	R_2	R_3	Class
Trichodermin	H	OAc	H	E
Verrucarol	H	OH	OH	E
15-Desacetyl-calonectrin	OAc	H	OH	E
3,15-Didesacetyl-calonectrin	OH	H	OH	E
Trichothecin (8-keto)	H	Oisocrot.	H	E
Trichothecolone (8-keto)	H	OH	H	E
Crotocin (7,8-epoxy)	H	Oisocrot.	H	E
Crotocol (7,8-epoxy)	H	OH	H	E
Scirpenetriol	OH	OH	OH	E-I_2
15-Acetoxy-scirpendiol	OH	OH	OAc	E-I_2
3-Desacetyl-calonectrin	OH	H	OAc	E-I_2

Initiation inhibitors

	R_1	R_2	R_3	R_4	R_5	Class
Verrucarin A, B. J	H	-MC-[a]		H	H	I_1
Roridin A	H	-MC-[a]		H	H	I_1
T-2 Toxin	OH	OAc	OAc	H	Oisoval	I_2
HT-2 Toxin	OH	OH	OAc	H	Oisoval	I_2
Neosolaniol	OH	OAc	OAc	H	OH	-
Diacetoxy-scirpenol	OH	OAc	OAc	H	H	I_2-E
Calonectrin	OAc	H	OAc	H	H	I_2-E
Nivalenol (8-Keto)	OH	OH	OH	OH	O	I_2
Fusarenon-X (8-Keto)	OH	OAc	OH	OH	O	I_2-E

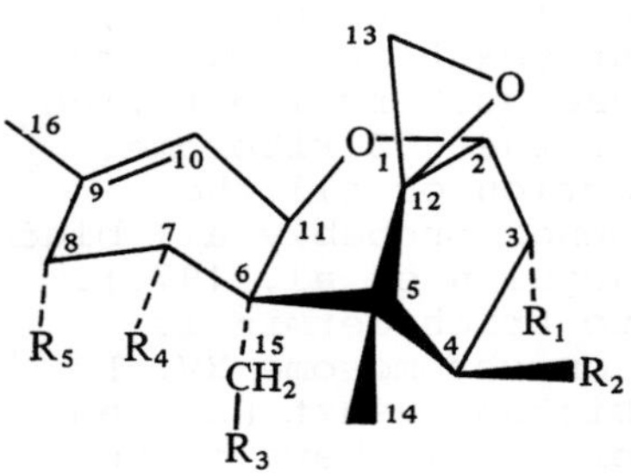

[a] MC=macrocyclic ring connecting R_2 and R_3.

In vivo studies confirm the primary mode of action of the trichothecenes at the level of protein synthesis. Oldham et al. (1980) studied the ultrastructure of cultured mouse fibroblasts treated with T-2 toxin and showed that the major effects of the toxin were on the rough endoplasmic reticulum which was poorly developed with few or no ribosomes present in T-2 toxin treated cells. Steele et al. (1981) studied the specificity of diacetoxyscirpenol, trichodermol and trichodermin as inhibitors of polypeptide initiation, elongation and termination in rat liver in vivo by analyzing polysome profiles. Diacetoxyscirpenol inhibited initiation at doses producing up to 70% inhibition of protein synthesis and both initiation and elongation at higher doses, confirming the in vitro data reported in Table 4. Both trichodermin and trichodermol were found to be inhibitors of elongation in vivo.

III. Interaction of Trichothecenes with Proteins

The first binding experiments involving trichothecenes was performed by Wei et al. (1974) who demonstrated that [acetyl-^{14}C] trichodermin bound to rabbit reticulocyte ribosomes in a reversible manner with a $K_a = 9.2 \times 10^5$ and with a maximum binding ratio of 0.44 molecules/ribosome. This binding is competitive with other trichothecene compounds (Cannon et al. 1976). [^{3}H]-Labeled T-2 toxin has been shown to bind to the albumin fraction of human serum in a reversible manner with a $K_a = 8.6 \times 10^4$ and a binding ratio of 0.11 T-2 toxins/serum albumin (Ito et al. 1979). Thus trichothecenes are able to bind to proteins and protein complexes in a readily reversible manner and can bring about their effects on protein synthesis without undergoing reactions involving covalent modification.

That several of the trichothecenes can undergo reaction with proteins resulting in a covalent modification was demonstrated by Ueno and Matsumoto (1975). These workers incubated fusarenon-X, neosolaniol and T-2 toxin with various thiol containing enzymes such as lactate dehydrogenase, alcohol dehydrogenase, and creatine phosphokinase, and demonstrated an enzyme inhibition which could be prevented by thiol protecting reagents such as dithiothreitol. [^{3}H]-Labeled fusarenon-X formed a 3.5:1 molar complex with alcohol dehydrogenase (a tetrameric protein) which could be isolated by gel filtration chromatography (Ueno and Matsumoto 1975). Incubation of dithiothreitol, cysteine or glutathione with fusarenon-X did not diminish the toxicity of the trichothecene to mice or its inhibition of protein synthesis, indicating that the compounds do not react with sulfhydryl groups indiscriminately. It may well be this ability to react with specific protein sulfhydryl groups which is responsible for the inhibition of mitosis observed in cultured cells of *Allium cepa* root tips treated with diacetoxyscirpenol (Reiss 1974) or Satratoxin H or T-2 toxin (Linnainmaa et al. 1979). Tubulin dimer, the major protein component in the microtubules of the mitotic spindle, has 14 free sulfhydryl groups (Kuriyama and Sakai 1974; Mellon and Rubhun 1976; Coss et al. 1981); titration of as few as four of these sulfhydryl groups completely blocks assembly of tubulin into microtubules (Kuriyama and Sakai 1974) and should result in the inhibition of mitosis in a manner similar to other mitotic spindle poisons (Wilson et al. 1974).

The apparent sparing of the liver from gross pathological changes which affect other tissues in animals given acute sublethal doses of trichothecenes (Lafarge-Fraysinnet et al. 1979) might be indicative of a specific liver detoxification mechanism for these toxins. Foster et al. (1975) reported that glutathione-S-transferase from liver catalyzes the formation of a glutathione-conjugate with trichothecenes such as T-2 toxin and diacetoxyscirpenol. However, Nakamura et al. (1977) were unable to reproduce these results.

IV. Mitochondrial Respiration

The effects of the trichothecene T-2 toxin on respiration of rat liver mitochondria was investigated by Bamburg (1969) and Kosuri et al. (1971). Concentrations of the toxin up to 1 mg/ml (2.1 mM) failed to inhibit respiration when glutamate was used as a substrate. Recently, however, two abstracts have appeared which reported an effect of T-2 toxin on respiration in rat liver mitochondria. Schiller and Yagen (1981) reported that T-2 toxin (2.5 mM) and T-2 tetraol (10 mM) inhibited state 3 respiration and prevented dinitrophenol stimulation with each of 4 substrate mixtures. Preincubation of the mitochondria with toxin for 5 min caused decreased respiratory control ratios. Pace and Murphy (1982) also investigated the effect of T-2 toxin (2.2 mM) on rat liver mitochondria and found a 40% inhibition in oxygen consumption in ADP-coupled mitochondria. Both site I and site III were affected by T-2 toxin. Following administration of an LD_{50} dose of T-2 toxin to rats, decreases of 12% in succinate oxidation and 45% in pyruvate oxidation were measured. Considering the fact that the doses needed to bring about partial inhibition of respiration of mitochondria in vitro are 3 to 6 orders of magnitude higher than those needed to kill eukaryotic cells in culture or to inhibit protein synthesis, the significance of the in vitro findings is of questionable relevance to the mode of action of these compounds. The results obtained in liver of animals given an LD_{50} dose of the T-2 toxin are of more significance but may arise from secondary effects due to inhibite protein synthesis and/or altered enzyme activities due to sulfhydryl reactivity directly with the T-2 toxin.

H. Physiological Mechanisms of Trichothecene Toxicity

While not all trichothecenes have been tested in every animal, it is probably safe to say that most trichothecenes are quite toxic to a wide variety of animals including poultry, mice, rats, guinea-pigs, rabbits, cats, pigs, cattle, monkeys and humans. Following ingestion of the trichothecenes, many animals show similar symptoms including diarrhea, vomiting, hyperemia of the intestine often followed by hemorrhaging and listlessness. Often neurological disorders also result from trichothecene ingestion. Topical application of the trichothecenes causes cutaneous irritation and necrotic lesions.

In reviewing the physiological effects of the trichothecenes in whole animals I will attempt to integrate what is known about the mechanism of the trichothecenes at the biochemical level with the symptoms observed at the organismal level. While this approach has some advantages in restricting the interpretation of whole animal data to known biochemical mechanisms, it must also be remembered that many toxins have specific pathways of activation in the whole animal and the complex symptoms of whole animal toxicity are often more than simply a sum of the effects of the toxin on different organ systems.

I. Mutagenic and Carcinogenic Activity

The first long term animal feeding study aimed at elucidating the carcinogenic potential of trichothecenes was performed by Marasas et al. (1969). Trout were fed a diet containing 200-400 ppb T-2 toxin for up to 12 months and no adverse effects, nor hepatoma development, were observed. Albino rats fed diets of 5-15 ppm T-2 toxin were stunted in growth and had inflammations of the skin around the mouth and nose; animals kept on a diet containing 10 ppm for 8 months (consumed 20 times the single LD_{50} of T-2 toxin) failed to show any significant pathological changes. Chronic feeding studies of fusarenon-X at 3.5 to 7 ppm in the diet of Donryu rats for 1-2 years also caused a slight decrease in weight gain, but tumor incidence was as low as the control group (Saito et al. 1980). Thus, it would appear that long term exposure to low doses of trichothecenes in the diet is not particularly harmful to laboratory animals.

Schoental et al. (1979) administered T-2 toxin intragastrically to white rats in doses of from 0.2 to 4 mg/kg body wt. (oral LD_{50} in white rats = 3.8 mg/kg, Kosuri et al. 1971). Animals received these doses 3 to 8 times, and in animals surviving 12 to 27.5 months (about 1/3 of the starting number) significant numbers of tumors and cardiovascular lesions were observed. High levels of cirulating T-2 toxin must have occurred following administration of these acutely toxic doses. Damage to the vascular endothelium, which resulted in hemorrhages in many remote organs including the brain, were observed. If the animal survived, repair of the endothelium occurred giving rise to thickened arterial walls and chronic cardiovascular lesions. Tumors of the liver were never observed but malignant or neoplastic changes were observed in squamous and glandular cells of the stomach, duodenum, pancreatic exocrine and islet cells, brain, pituitary, and adrenal medulla. While these findings are striking and might indicate that T-2 toxin is indeed a carcinogen when in circulation, the potent radiomimetic effects of T-2 toxin, especially on the cells of the lymphoid system (Section H.III.), might also explain the high incidence of tumors in T-2 toxin treated animals.

T-2 toxin failed to induce any papillomas when applied topically to mice even if the applications were followed with the promoter, croton oil (Marasas et al. 1969; Lindenfelser et al. 1974). A few papillomas developed on mice that had been treated with 7, 12-dimethylbenz [a] anthracene followed by T-2 toxin (Marasas et al. 1969), although the numbers of such tumors were not statistically significant (Lindenfelser et al. 1974). Lafarge-Frayssinet et al. (1981) reported the development of leukemia in a mouse which received T-2 toxin painted on its skin over an unspecified period of time. Although T-2 toxin may be a weak cocarcinogen in the papilloma induction experiments, it may also allow tumors to develop by suppressing the animal's immune system.

Several tests for mutagenicity have been carried out with the trichothecenes and, to date, all have been negative for the class A and B compounds. Nagao et al. (1976) and Ueno (1977) have tested diacetoxyscirpenol, fusarenon-X and T-2 toxin in the *Salmonella typhimurium* mutagenicity assay (Ames test – Ames et al. (1973)). Doses of 0.01-500 μg per plate were used and all these trichothecenes were negative with or without the "activation" by liver homogenate fraction S-9 (Ames et al. 1973). Ueno (1977) did report that crotocin, a diepoxide-containing class D trichothecene, had mutagenic activity, but only at 500 μg/plate, the highest dose tested. Ueno et al. (1978), Kuczuk et al. (1978), Pääkkonen et al. 1978 and Wehner et al. (1978) have all used the *S. typhimurium* mutagenicity assay with several other trichothecenes including Satratoxin H of the class C macrocyclic trichothecenes and, with the exception of the diepoxy class D trichothecenes, all gave negative results for up to 500 μg/plate, the highest doses tested.

Other tests for the mutagenic potential of trichothecenes have included the sex-linked recessive lethal test of *Drosophila* (Sorsa et al. 1980). Both T-2 toxin and the macrocyclic trichothecene Satratoxin H were tested in this assay and both were non-mutagenic, although slight increases in maternal and paternal sex chromosomal non-disjunction were observed in offspring of flies fed these mycotoxins.

In higher animals the mutagenic potential of T-2 toxin was studied by its ability to induce chromosomal aberrations in bone marrow cells of Chinese hamsters (Norppa et al. 1980). At acute but sublethal concentrations (~3 mg/kg body wt.), T-2 toxin induced some rare types of exchange aberrations in metaphase cells of the bone marrow, indicating that chromosome damage can occur following intraperitoneal administration of high, but sublethal doses. Oral administration of 2.5 mg/kg T-2 toxin once a week for six weeks failed to induce any significant chromosomal aberrations.

Stahelin et al. (1968) first reported the reduction in mitotic index of cultured animal cells by the trichothecene diacetoxyscirpenol. Similar results have been reported for plant cell (*Allium cepa* root tip) cultures (Reiss 1974) in which diacetoxyscirpenol, at concentrations of 0.1 μg-1000 μg/ml, induced chromosomal aberrations. Linnainmaa et al. (1979) did not observe any chromosomal aberrations in *Allium cepa* root meristem cultures following treatment with 20 or 100 μg/ml T-2 toxin or 10 μg/ml satratoxin H. However, both toxins were strong mitotic inhibitors, and they arrested mitosis in a manner comparable to the action of colchicine. Animal cells are much more sensitive to the cytostatic effects of the trichothecenes than are plant cell cultures. Dosik et al. (1978) reported that diacetoxyscirpenol (anguidine) at concentrations as low as 0.1 μg/ml effectively induced a "frozen cell cycle state", (i.e., no progression through any phase of the cycle) in cultured human colon cancer cells (LoVo cells). More recently, Teodori et al. (1981) have shown that this "frozen cell cycle state", induced by diacetoxyscirpenol in cultured Chinese hamster ovary cells, protects these cells from the cytotoxic effects of adriamycin and 1-β-D-arabinofuranosylcytosine, drugs which rely on cell cycle progression for killing. Flow cytometry studies of cells treated with diacetoxyscirpenol alone showed that within 3 h after exposure to the trichothecene slight redistribution of the cells within the cell cycle took place with a 10-20% increase occurring in cells in S phase and a 5-10% decrease in cells in G_1 and G_2+M. The distribution of cells in the cell cycle observed at 3 h remained relatively constant during continuous exposure to diacetoxyscirpenol.

Although the term "frozen cell cycle" might imply that the trichothecenes demonstrated a reversible proliferative inhibiting effect on cultured cells, studies with T-2 toxin demonstrated that the cloning efficiency of cultured normal human fibroblasts was much more sensitive to the toxin than was cell survival (Oldham et al. 1980). Thus the levels of the trichothecenes needed for cell cycle arrest (0.1 μg/ml for diacetoxyscirpenol), which might be reversible in terms of cell survival, may destroy the proliferative capacity (cloning capability) of the cells. Oldham et al. (1980) demonstrated that the LD_{50} for cultured fibroblasts treated with T-2 toxin was 0.7 μg/ml, whereas 0.004 μg/ml of T-2 toxin reduced the cloning efficiency of the cells (ability to form colonies of $\geq$ 100 cells) by 50% (ED_{50}). The ED_{50} dose had very little effect on either the rate of protein synthesis or DNA synthesis in these cultured fibroblasts, whereas the LD_{50} dose resulted in almost a complete inhibition of both of these macromolecular synthetic processes within one hour.

In order to determine if the effect of the low concentrations of T-2 toxin on the reduced proliferative capability of normal human fibroblasts was due to DNA damage, Oldham et al. (1980) and Agrelo and Schoental (1980) studied the amounts of unscheduled DNA synthesis (DNA repair) in cultured cells treated with both T-2 toxin and hydroxyurea (an inhibitor of scheduled DNA synthesis). Agrelo and Schoental (1980) failed to see a stimulation of unscheduled DNA synthesis when cells were treated with either T-2 toxin or HT-2 toxin at concentrations ranging from 0.006-100 μg/ml. However, these workers did report a significant increase in unscheduled DNA synthesis in cells treated with HT-2 toxin which had been treated with rat liver microsomal fractions S-9 (Ames et al. 1973), an indication that metabolic activation of the trichothecenes might be necessary to convert the compounds to a form capable of damaging DNA. However, even in the presence of the microsomal fraction, HT-2 toxin only had significant stimulatory activity on unscheduled DNA synthesis at 100 µg/ml. Oldham et al. (1980) used a more sensitive autoradiograph procedure for measuring unscheduled DNA synthesis in fibroblasts. These workers reported that unscheduled DNA synthesis took place over a wide range of concentrations for both T-2 toxin and T-2 tetraol starting at toxin concentrations equivalent to those which inhibited the proliferative capacity of the cells. These findings indicate that the reduced proliferative capacity of cells exposed to low levels of trichothecenes may result from DNA damage.

More direct evidence for DNA damage by low levels of T-2 toxin has been reported by Lafarge-Frayssinet et al. (1981). These workers studied the induction of single strand breaks in DNA of cultured lymphocytes and hepatocytes by 5 ng/ml of T-2 toxin for short exposure times (2 h). Single strand breaks were also evaluated in DNA of liver, spleen and thymus glands of mice 3 h after a single intraperitoneal administration of 3 mg/kg body wt. of T-2 toxin. In both cultured cells and animal organs no damage was observed in hepatic DNA, whereas severe damage occurred to the DNA of lymphoid cells and organs. These findings on the preferential action of T-2 toxin on the lymphoid organs is in agreement with previous reports (Saito et al. 1969; Sato et al. 1975; Schoental and Joffe 1974; Lafarge-Frayssinet et al. 1979) and, although they are suggestive of a compound with carcinogenic potentiality, they also suggest that tumor development in animals treated with trichothecenes may result from the reduced immunological capability of the animal. The DNA damage in lymphoid tissue may in fact arise indirectly through the action of the trichothecenes on lysosomes which contain increased amounts of hydrolytic enzymes in the lymphatic tissue (Pokrovsky et al. 1976).

II. Teratogenic Effects

Injection of labeled T-2 toxin into pregnant rats demonstrated that this trichothecene readily crosses the placenta; the toxin was found in highest amounts in the macromolecular fraction of the thymus gland of the newborn where it seemed to accumulate (Lafarge-Frayssinnet et al. 1980). It is not surprising, therefore, to learn that intraperitoneal injections of T-2 toxin at 0.5 mg/kg body wt. into pregnant mice on day 10 of gestation induced tail and limb anomalies (Stanford et al. 1975). The teratogenic effects of T-2 toxin could be significantly enhanced if ochratoxin were also present (Hood et al. 1978). Ochratoxin is itself a known teratogen (Hayes et al. 1974), but the fetal malformations observed when it was administered along with T-2 toxin are strictly those of the T-2 toxin type.

Studies carried out on laying hens indicate that T-2 toxin at a level of 20 ppm in the chicken feed lowered egg production by about 20% and resulted in eggs with significantly thinner shells (Wyatt et al. 1975a). Similar findings were reported by Chi et al. (1977a) who administered 8 ppm T-2 toxin in the feed; Chi et al. also quantitated the hatchability of the eggs and found over a 10% reduction in viability over an 8 week feeding period. Death of the embryo occurred during the early stages of development but no evidence of malformation in either the dead or hatched chicks was found in the T-2 toxin treated groups (Chi et al. 1977a). Newly hatched chicks fed a diet containing T-2 toxin at a level of 4-16 ppm in the feed for 3 weeks showed growth inhibition and abnormal feathering over the entire body of the chicken (Wyatt et al. 1975b).

III. Immunosuppressive Activity

The trichothecene mycotoxins have been described as having a "radiomimetic" effect on animals (Saito et al. 1969), resulting in marked damage to the proliferating cells of the hematopoietic tissue in the bone marrow, spleen, thymus and lymph nodes (Saito and Tatsuno 1971). This damage to the lymphoid organs probably results from the inhibitory action of the trichothecenes on protein synthesis; however, the fact that the damage to the lymphatic tissue is preferential over other tissues and the finding of DNA damage exclusively in lymphoid cells (Lafarge-Frayssinet et al. 1981) indicate that many toxicological symptoms may arise from this immunospecific form of action.

The major effects of trichothecenes on the necrosis of the lymphoid tissues occur during acute toxicity and are not as pronounced during chronic feeding studies. These effects have been observed in guinea-pigs given oral doses of diacetoxyscirpenol (Kriegleder 1981) and in guinea-pigs (DeNicola et al. 1978), turkeys and chickens (Richard et al. 1978), mice (Hayes et al. 1980; Lafarge-Frayssinet et al. 1981), cats (Sato et al. 1975) and monkeys (Rukmini et al. 1980; Jagadeesan et al. 1982) administered T-2 toxin. Fusarenon-X (Ueno et al. 1971b) and nivalenol (Saito and Tatsuno 1971) have also been well characterized as causing necrosis of the lymphatic tissues. The immunosuppressive effects in mice of small doses of diacetoxyscirpenol made the animals sensitive to fungal infections. The mice succumbed to doses of *Candida albicans* (Fromentin et al. 1980) and *Cryptococcus neoformans* (Fromentin et al. 1981) which were insufficient to kill animals that had not been treated with the trichothecene.

More specific investigations into the nature of the immune deficiency have been performed. Masuko et al. (1977) and Otokawa et al. (1979) studied the effect of T-2 toxin on delayed hypersensitivity in mice following sensitization by sheep red blood cells (SRBC). The delayed hypersensitization response was enhanced only when T-2 toxin was administered several days after the sensitization and no effects were observed on the formation of antibody. T-2 toxin treatment caused a marked decline in the cell population of the thymus. Based on their observations the authors suggested that the trichothecene was interfering with the generation of suppressor cells involved in the delayed hypersensitivity response.

Rosenstein et al. (1979) showed that mice treated via intraperitoneal injection with T-toxin, diacetoxyscirpenol or a crude extract of *Fusarium poae* for 3 days before sensitization by SRBC showed a marked decrease in anti-SRBC production and a decline in thymus weight. The decline in antibody production occurred at doses of the trichothecenes which had little effect on the weight of the thymus but T-2 toxin was more effective than diacetoxyscirpenol in reducing both of these parameters. Mice immunized during the course of T-2 toxin treatment have a very low anti-SRBC titre, but when T-2 toxin treatment ended the antibody titre returned to the level found in immunized controls in about 6 days. Administration of T-2 toxin to mice 3 days after immunization with SRBC resulted in the antibody titre rapidly reaching a plateau; the titre continued to increase in immunized control animals. Rosenstein et al. (1979) also demonstrated that rejection of skin allografts was significantly delayed in animals receiving T-2 toxin indicating a decrease in the cell-mediated immune response. Decreased numbers of macrophages and lymphocytes were observed in the region of the allograft.

Rosenstein et al. (1981) reported that spleen cells from mice administered T-2 toxin, diacetoxyscirpenol or a crude extract of *F. poae* demonstrated an enchanced response to antigen which indicates that T-independent immune responses can be increased by trichothecene treatment. While the mechanism of this complex response to trichothecenes is not altogether clear, Rosenstein et al. (1981) believe that a selected effect of these toxins on subpopulations of T-suppressor cells or their precursors might explain these findings.

The effects of T-2 toxin on the immune system have also been studied in larger animals. T-2 toxin was administered orally to monkeys (Jagadeesan et al. 1982) and calves (Mann et al. 1982). Monkeys received 0.1 mg/kg/day for 4-5 weeks, while the calves received 0.6 mg/kg/day for 43 days. The monkeys showed a significant decrease in both T-cell and B-cell numbers and a very large decrease in plasma immunoglobulin levels, especially IgG. T-cell stimulation by phytohemaglutinen was also suppressed. In the bovine serum, levels of IgG did not change but IgM and IgA levels were markedly reduced. The total serum protein in T-2 toxin treated calves was about 10-15% less than in control animals on an equivalent feed intake. Calves on the T-2 toxin-containing diet were immunized with anaplasma on days 1 and 21 of the feeding study (Mann et al. 1982). The primary response to the antigen was depressed (an IgM response), whereas the secondary or memory response was normal (IgG response).

Further studies on the mechanism of the immunosuppressive effect of the trichothecene fusarenon-X in mice have recently been reported (Masuda et al. 1982). The mitogenic response of mouse splenic lymphocytes was suppressed by pretreatment in vitro with 50 ng/ml or higher concentrations of fusarenon-X; more than 90% of the cells remained viable when up to 5 μg/ml of fusarenon-X was used. Lymphocytes prepared from mice treated in vivo with fusarenon-X (25-50 μg/day) also showed a marked decrease in mitogenic response. The inhibition was more pronounced to T-cell mitogens (such as Concanavalin A) than to B-cell mitogens (such as a bacterial lipopolysaccharide). Both the IgE and IgG1 antibody responses were depressed when fusarenon-X was given for 7 days before immunization. Thus both T-lymphocyte function and antibody production were depressed, again suggesting that it is the dysfunction of the helper T-lymphocytes which results in the decreased antibody titre.

IV. Hematological Effects

The effects of several trichothecenes on hematological parameters have been studied in mice (Hayes et al. 1980), guinea-pigs (DeNicola et al. 1978), rabbits (Rüsch and Stähelin 1965; Gentry and Cooper 1981), cats (Sato et al. 1975; Lutsky et al. 1978), dogs (Rüsch and Stähelin 1965), chickens (Joffe and Yagen 1978; Richard et al. 1978; Pearson 1978; Chi et al. 1981; Chi et al. 1977b; Doerr et al. 1981), pigs (Rüsch and Stähelin 1965; Weaver et al. 1978b, 1978d, 1981; Patterson et al. 1979), calves (Patterson et al. 1979; Mann et al. 1982), monkeys (Rüsch and Stähelin 1965; Jagadeesan et al. 1982) and man (Goodwin et al. 1978).

In the small mammals, oral administration of the toxins at levels of 0.1 mg - 1 mg/kg body wt./day (or up to 20 ppm in the diet) gave similar results, although animal sensitivity to the toxins varied and the severity of the changes often depended on the method of administration. Following a brief initial period of leukocytosis, the white blood cell count dropped rapidly to 10-20% of controls with a smaller but significant reduction seen in red cells as well. A period of anemia in mice was followed by regeneration of the erythropoietic system while the animals were still consuming a diet of 20 ppm T-2 toxin (Hayes et al. 1980). Mice have a shorter erythrocyte life than other animals and so the anemia appeared more rapidly. Most animals also showed an increase in neutrophiles. When the trichothecenes were removed from the diets, all animals recovered and blood parameters returned to control levels.

Monkeys appeared to be quite sensitive to 0.1 mg/kg T-2 toxin/day (Rukmini et al. 1980; Jagadessan et al. 1982). Complement levels were near normal, but a severe leucopenia resulted affecting both T-cell and B-cell populations. IgG levels decreased substantially and IgM levels decreased slightly after 4-5 weeks of toxin administration.

Pigs fed diets containing 10 ppm diacetoxyscirpenol showed no changes in blood parameters including packed cell volume, hemoglobin levels, red cell and white cell counts, total protein, blood urea nitrogen, and blood enzyme levels. After recovery from an acute dose (up to 3.2 mg/kg body wt.) of T-2 toxin, pigs had no abnormalities in their blood parameters, (Weaver et al. 1978b). Administration of 0.2 mg/kg T-2 toxin or diacetoxyscirpenol to calves for 11 days caused a slight increase in prothrombin times and in partial thromboplastin times, but these effects were not observed in pigs (Patterson et al. 1979). Administration of higher doses (0.6 mg/kg/day) of T-2 toxin for 6 weeks resulted in a decreased serum protein level in calves especially in the globulin fraction, whereas albumin levels were slightly elevated (Mann et al. 1982). Immunoglobulins A and M were decreased, whereas IgG levels were near normal after 43 days of toxin administration. The serum levels of C3 dropped by 30% during the same time period.

Chickens fed up to 4 ppm of T-2 toxin in their diet showed no significant changes in hematological parameters over a period of 9 weeks except for a decrease in serum alkaline phosphatase (Chi et al. 1977b). At levels of T-2 toxin in the diet of 4 ppm-16 ppm a qualitative change was demonstrated in the lipid fraction; thromboplastin and Factor VII activity was significantly reduced (Doerr et al. 1981). Factor X, prothrombin and fibrinogen levels were all depressed by 16 ppm T-2 toxin in the diet with the latter two parameters reduced below 60% of normal. The blood clotting inhibitory effects only occurred at much higher T-2 toxin levels than the 0.5 ppm of T-2 toxin which caused oral lesions in the chick (Wyatt et al. 1972b).

Diacetoxyscirpenol (also called anguidine or NSC-141537) has been tested in phase I and phase II clinical trials as an anticancer agent in humans. In phase I trials a toxicological evaluation was carried out over a 5 day period (Goodwin et al. 1978). An infusion of 4.5 mg/m^2 (about 120 μg/kg body wt.) over a 4-8 h period was used. Leukopenia and thrombocytopenia ranging from mild to severe occurred in over half the patients given the drug. No significant changes in blood chemistry parameters including blood urea nitrogen, creatine, serum aminotransferases, alkaline phosphatase, bilirubin, immunoglobulin, complement, prothrombin time, fibrinogen levels, etc. were observed.

When white rats were treated with 2 mg/kg of T-2 toxin by stomach tube at weaning and again at 3 months followed by 1 mg/kg at 6 months and 3 mg/kg at 12 months, their blood pressure was found to increase between 13 and 17 months after the first treatment (Wilson et al. 1982). Cardiovascular lesions were observed in these rats with the arterial endothelium undergoing reparative processes which led to a thickening which partly occluded the lumen. Calcification of the vessel walls was also observed in some animals.

Based on the above results it appears that similar effects on hematology in different animal species can be achieved at different doses of trichothecenes. The method of toxin administration as well as the regimen of doses and the time after treatment when the parameters are evaluated, all have an impact on the nature of the changes which will be observed.

V. Neurological Disorders

Many of the naturally occurring mycotoxicoses in which trichothecenes have been implicated as the causative agent are characterized by neurological manifestations, e.g. "staggering grain toxicosis" (Woronin 1891) and alimentary toxic aleukia (Mayer 1953). Neurological disturbances observed in chickens following ingestion of dietary T-2 toxin include abnormal positioning of the wings, hysteriod seizures, and an impaired righting reflex in young birds (Wyatt et al. 1973). On a diet containing 16 ppm T-2 toxin birds started showing these symptoms by day 13, whereas 17 days were required to see these symptoms when the diet contained 8 ppm T-2 toxin. Diets containing as little as 4 ppm T-2 toxin, could induce these symptoms after 3 weeks. Chi et al. (1981) administered T-2 toxin to four-week old chickens in a single dose of 2.5 mg/kg body weight and then followed the brain concentrations of dopamine, norepinephrine and serotonin over the next 48 hours. While serotonin levels were not altered from those of controls, 25 hours after T-2 toxin administration brain dompamine levels were almost double the control while norepinephrine levels had fallen by 25%. Since dopamine is a precursor to norepinephrine, it is possible that the toxin has some direct inhibitory effect on this conversion. Changes in brain catecholamine levels certainly could be responsible for altered motor activity.

Since dopamine is a precursor to norepinephrine, it is possible that the toxin has some direct inhibitory effect on this conversion. Changes in brain catecholamine levels certainly could be responsible for altered motor activity.

Administration of diacetoxyscirpenol (anguidine) to humans resulted in several neurological symptoms (Goodwin et al. 1978). Administration of 3.1-6.0 mg/m^2 (about 80-150 µg/kg body wt.) five times per day caused abnormal EEG's in several patients; neurological symptoms ranged from mild confusion to short periods of coma and hallucination. Headaches occurred in 19% of the patients. In a separate study (phase II clinical trials) about 12% of the 93 patients treated with infusion doses of 3-4.5 mg/m^2 (80-120 µg/kg body wt.) diacetoxyscirpenol/day for 5 days reported hallucinations and/or confusion (Bukowski et al. 1982). Thus there is no doubt that trichothecenes at relatively low concentrations in humans can induce the neuropsychiatric manifestations which have been described for the outbreak of alimentary toxic aleukia (Mayer 1953). It is likely that altered catecholamine levels in human brain also occur in response to trichothecene administration and bring about the hallucinogenic symptoms.

Cannon et al. (1982) recently demonstrated that intraventricular injection of 3,15-didesacetylcalonection (60 µg) or T-2 toxin (10 µg) caused a significant decline in protein synthesis in the hypothalamus of the rabbit. Blocking protein synthesis in the hypothalamus also prevented the fever response to injections of leukocyte pyrogen which indicates that protein synthesis within the brain may be a necessary step in the activation of central pyrogenic mechanisms. As a control, these researchers used a modified calonectrin which had a reduced 12,13-epoxide group and which was inactive in inhibiting protein synthesis in vitro (Carter and Cannon 1977). This derivative was also unable to inhibit the pyrogenic response. Normal thermoregulation against cold was not inhibited by blockage of hypothalamic protein synthesis. Thus, high levels of trichothecenes in the diet of animals which cause immunosuppression and make animals more susceptible to secondary infections, might also suppress the fever ordinarily brought about by these infections.

VI. Vomiting

In many animals including cats (Sato et al. 1975), dogs (Matsuoka et al. 1979), pigs (Vesonder et al. 1973), ducklings (Ueno et al. 1974) and man (Goodwin et al. 1978; Bukowski et al. 1982), vomiting is one of the earliest symptoms of trichothecene ingestion. Ueno (1980b) has compared minimum doses of trichothecenes which induce vomiting in several different animals and, as a general rule, the class A trichothecenes are more active than those with an 8-keto group (class B). Deoxynivalenol, a trichothecene isolated because of its ability to induce vomiting, was given the trivial name vomitoxin (Vesonder et al. 1973) but its minimum effective dose in inducing vomiting in ducklings is 100 times greater than T-2 toxin or neosolaniol. However, deoxynivalenol is much more active in inducing vomiting in cats, dogs and swine than it is in the ducklings (Ueno 1980b). Nausea and vomiting occurred in 10 out of 10 of the human subjects given 3.1 to 4.5 mg/m^2 (80-120 µg/kg body wt.) of diacetoxyscirpenol by rapid intravenous injection during phase I clinical trials (Goodwin et al. 1978) and in 49% of 93 patients given similar doses of the drug by slow (4-8 h) infusion (Bukowski et al. 1982).

Studies by Matsouka et al. (1979) showed that pretreatment of dogs with metoclopramide hydrochloride or chlorpromazine hydrochloride 1 h before injection with fusaranon-X reduced the incidents of vomiting and significantly lengthened the latency period between vomiting. Based upon these results the authors suggest that fusaranon-X may work directly on a chemoreceptor trigger zone in the medulla oblongata which stimulates vomition.

VII. Diarrhea

Diarrhea is a common symptom in animals administered trichothecenes regardless of the route of administration. Matsuoka and Kubota (1981, 1982) have studied the mechanism by which fusarenon-X induces diarrhea in rats; intraperitoneal administration was used to avoid direct effects on the intestinal mucosa. Following intraperitoneal injection of 1 mg/kg body wt. of fusarenon-X, watery diarrhea usually occurred within 36 to 60 hours. Autopsies done 24 h after injection revealed expansion of the small intestine and leakage of intravenously injected Evans blue dye into the intestinal lumen. The average length of intestinal villi was shortened and an infiltration of erythrocytes into the intestinal lamina propria was observed. Therefore, increased permeability of the blood vessel walls and the intestinal mucosal epithelium must occur in response to the trichothecene, leading to a leakage of plasma contents into the intestinal lumen. This extravization of circulatory fluid is also observed following topical application of trichothecenes. Whether or not severe hemorrhaging will occur probably depends on the dose of the trichothecene administered as well as the nutritional state of the animal receiving the toxin. Vitamin deficiencies are often characterized themselves by capillary fragility (Chaney 1982) and may also prevent proper functioning of some enzyme catalyzed detoxification processes which can exacerbate the toxicity of the compounds (Schoental 1980a).

The direct effects of single oral doses of T-2 toxin (2.5 or 5 mg/kg) on the digestive tract of the guinea-pig were observed by DeNicola et al. (1978). Mucosal lesions were present throughout the gastrointestinal tract but were most severe in the caecum and stomach. The alterations in the caecum consisted of mucosal hemorrhage, necrosis of the mucosal epithelium and necrosis within the centers of the subepithelial lymphoid nodules. The fundic mucosa of the stomach was most severely involved with mucosal hemorrhage and necrosis also apparent. In the small intestine focal areas of villar necrosis and cast formation within crypts were found.

Matsouka and Kubota (1982) also studied the absorption rate of D-xylose and D-glucose using intestinal sacs prepared from control rats or from rats treated 24 h before sacrifice with fusarenon-X (1 mg/kg) administered by intraperitoneal injection. Sugar transport was altered in the trichothecene treated animal. An abnormal increase in permeability was found and sugar translocation seemed to depend on simple diffusion implying that the transport functions of the intestinal mucosal cell membranes had been altered by trichothecene treatment of the animal.

VIII. Hemorrhagic Syndrome

Trichothecene mycotoxins, most notably T-2 toxin and diacetoxyscirpenol have been identified in, or isolated from moldy feed implicated in moldy corn toxicosis and hemorrhagic syndrome in cattle and pigs (Hsu et al. 1972; Kurtz et al. 1976; Petrie et al. 1977). Huff et al. (1981) reported that large doses of deoxynivalenol administered to day old chicks via oral intubation caused extensive ecchymotic hemorrhaging throughout the body along with other symptoms consistent with the description of hemorrhagic anemia snydrome in chickens caused by moldy feed (Forgacs and Carll 1962). Other workers, however, have reported that oral administration of large doses (0.1-0.6 mg/kg) of purified T-2 toxin or diacetoxyscirpenol to cows, calves or pigs failed to induce any evidence of hemorrhagic syndrome (Matthews et al. 1977, 1979; Weaver et al. 1980). This failure to obtain hemorrhaging probably does not eliminate the trichothecenes from the etiology of the hemorrhagic syndrome but rather indicates that the syndrome is of a far more complicated nature than having a single cause. The extravization of plasma proteins and red cells from capillaries observed following fusarenon-X treatment of rats indicates the potential for hemorrhage, but other dietary factors almost certainly are involved. The slight increase in prothrombin and partial thromboplastin times observed in T-2 toxin treated animals (Patterson et al. 1979), probably arise through partial inhibition of liver protein synthesis, and may also make the animals more prone to hemorrphage.

IX. Cutaneous Irritation

Most of the trichothecene compounds are potent skin irritants and inflammatory agents (see reviews by Bamburg and Strong 1971, and Ueno 1980a). Marasas et al. (1969) described the lesions induced on rat skin by 1.2 mg of T-2 toxin applied in ethyl acetate. Extensive inflammation and coagulation necrosis of the dermis occurred in 48-72 h followed by sloughing of the covering epidermis. Polymorphonuclear leukocytes had infiltrated tissue adjacent to the areas of coagulation necrosis. Tissue damage extended into the underlying subcutical area. Lower doses of 120 to 450 μg of T-2 toxin also resulted in epidermal necrosis. Ueno et al. (1970b) compared the skin necrotizing activity in different animals of diacetoxyscirpenol, fusarenon-X and nivalenol dissolved in acetone. The guinea-pig proved to be the most sensitive with doses of 100 μg of diacetoxyscirpenol and fusarenon-X producing very heavy necrotic lesions. Nivalenol was less active at the same concentration.

Hayes and Schiefer (1979) examined the cutaneous irritation by T-2 toxin and diacetoxyscirpenol on rats and rabbits. Histopathological observations were made from 12-48 h with findings similar to those reported by Ueno et al. (1970b) and Marasas et al. (1969). The reaction was comparable to that caused by croton oil, a tumor promoter. Of added interest is the fact that these authors found the trichothecenes to have greater necrotizing activity when applied in ethyl acetate or methanol than when applied in dimethylsulfoxide. As has been observed previously, the inflammation and edema seem to be brought on by the effects of the trichothecenes on increased capillary permeability in the treated area.

An irreversible depigmentation of the dark hair on a C57BL mouse was observed following topical application of T-2 toxin (Schoental et al. 1978). Depigmentation actually occurred at lower concentrations of the toxin than were needed for causing necrosis of the skin. Thus T-2 toxin has long term effects on the pathway for synthesis of melanins in the hair follicles of mice.

I. Metabolism of Trichothecenes

I. Toxicokinetics

Intravenous administration of T-2 toxin to pigs at doses from 0.3 to 0.6 mg/kg resulted in detectable levels of the toxin in plasma for relatively short time periods with the half-life estimated to be less than 5 min (Vesonder et al. 1981c). A similar half-life for T-2 toxin in the plasma of a calf following intravenous administration was also reported (Beasley et al. 1981). However, longer lived metabolites of T-2 toxin were identified in both pig and calf plasma as well as urine, liver and kidney tissue.

Oral administration to a lactating cow of T-2 toxin radiolabeled by the method of Wallace et al. (1977) was performed in order to follow the metabolic fate of the toxin in various tissue and body fluids (Yoshizawa et al. 1981). Plasma samples taken after oral administration of 156.9 mg T-2 toxin (408 µCi/mg – equivalent to 0.171 mCi/kg body wt.) contained a measurable quantity of T-2 toxin although more polar metabolites of the toxin were evident in higher concentration than the T-2 toxin itself. The level of T-2 toxin peaked 4 h after administration at 8 ppb, but it represented less than 2% of the radioactive metabolites in plasma at this stage. Plasma levels of T-2 toxin had dropped below 1 ppb by 20 h. Several more polar metabolites of T-2 toxin were also identified; one of these, referred to as TC-3, remained at levels of about 8 ppb for relatively long periods (12-16 h) after treatment. Only small amounts of the known metabolites HT-2 toxin and neosolaniol were observed. The radioactivity in plasma which peaked at 8 h post administration declined after that time with a half-life of 16 h. All of the major metabolites appeared to be transformation products of the cow, since none of these metabolites was detected in the rumen. The major metabolites, designated TC-1 and TC-6, in addition to the TC-3, do not correspond to any partial deacylation products of T-2 toxin in their chromatographic behavior. In addition, TC-1 and TC-3 both have intact 12,13-epoxide rings since they give a positive color reaction with 4-(p-nitrobenzyl)pyridine/tetraethylenepentamine reagent (Takitani et al. 1979).

Blood samples taken from Khmer Rouge guerrillas and other inhabitants of Tuol Chrey, Kampuchea 24 h after an attack of "yellow rain" were analyzed for trichothecenes using gas chromatography/mass spectrometry. Results showed T-2 toxin levels of 18 ppb in one victim and 22 ppb in another; HT-2 toxin levels in the same two individuals were 11 ppb and 10 ppb respectively. Levels of the toxin were detectable in blood samples removed from five other victims 18 days after the attack (Penn 1982, Interview). Assuming that the total toxin level (T-2+HT-2) in the victims' blood 24 h after exposure is 30 ppb and that the toxicokinetics in humans is roughly equivalent to the cow, the victims of the "yellow rain" attack absorbed the equivalent of 12.5 mg of toxin per kg body wt. However, it is certainly possible that the toxin could persist for longer periods in human plasma than in the cow and that more of it stays as T-2 toxin rather than the multiple metabolites found in the cow (Yoshizawa et al. 1981). Much lower exposure levels could then account for the T-2 toxin levels found in the blood samples.

II. Routes of Excretion

Ueno et al. (1971b) and Matsumoto et al. (1978) followed the metabolic fate of [^{3}H]-labeled fusarenon-X and T-2 toxin injected into mice and rats, respectively. About 50% of the total dose administered was found in the feces and 10% in the urine during the first 24 h. A small amount of nivalenol was detected in extracts of the excreta from fusarenon-X treated mice while T-2 toxin, HT-toxin and T-2 tetraol were identified in extracts of excreta from T-2 toxin treated rats.

The levels of radioactivity in feces, urine and milk following an oral dose of [^{3}H]-T-2 toxin to a lactating cow have been reported (Yoshizawa et al. 1981). The maximum levels of radioactivity were reached at the following times after dosing: feces at 44 h (9.2 ppm); urine at 16 h (5.5 ppm); milk at 16 h (37 ppb). By 72 h, 29% of the total radioactivity had been eliminated in the urine and 72% in the feces with only 0.2% being present in the milk.

Tritium labeled T-2 toxin has also been administered by oral intubation to pigs (Robison et al. 1979a) and chickens (Chi et al. 1978a). Weanling pigs died 18 h after administration of either 0.1 mg/kg or 0.4 mg/kg of T-2 toxin. About 20% of the radioactivity was found in the urine in both animals. Contents of the gastrointestinal tract were not assayed for radioactivity but feces from the low dose animal contained 25% of the administered radioactivity. Feces from the high dose animals contained less than 1% of the administered radioactivity indicating that a much larger fraction was still retained in the gastrointestinal tract. The bile of both animals contained a relatively high specific activity (dpm/mg tissue), indicating that a significant portion of the radioactivity which eventually is passed in the feces is first absorbed by the animal and processed by the liver. Kosuri et al. (1971) suggested that T-2 toxin was excreted as glucuronide conjugates based on the increased glucuronide production in animals administered T-2 toxin; no glucuoronide conjugates of trichothecenes have been identified, however.

Chickens excreted 81.6% of the radioactivity from a single oral dose of [^{3}H]-T-2 toxin (0.5 mg/kg) within 48 h post administration (Chi et al. 1978a). Another 10.4% of the radioactivity was present in the gastrointestinal tract. Of the 8% of the total radioactivity which was found in the chicken at 48 h, the bile and gall bladder contained by far the highest specific activity and about the same total amount as the muscle tissue (~3%). These findings also suggest that the liver plays a major role in the metabolism and excretion of the toxin and/or its metabolites into the feces via the bile.

III. Tissue Distribution

Other than in the bile and gall bladder, no specific tissue accumulation of labeled T-2 toxin or its metabolites was found in either rats (Matsumoto et al. 1978), chickens (Chi et al. 1978a), pigs (Robison et al. 1979a) or cows (Yoshizawa et al. 1981). Tissues that were tested for radioactive residue included muscle, heart, fat, spleen, liver, kidney, gall bladder and, in the chicken, the gizzard. Brain was not analyzed in any of the studies except as being included with the remainder of the carcass.

IV. Transmission in the Food Chain

In most industrialized societies farm animals are much more likely to consume moldy feed containing trichothecenes than are humans; therefore several studies have been done in order to assess the potential for human consumption of trichothecenes in the food chain. A lactating cow fed a single dose of [^{3}H]-labeled T-2 toxin which would have amounted to a level of T-2 toxin in the feed of over 30 ppm (higher than has been reported for any natural outbreak of T-2 toxicosis) had 0.2% of the T-2 toxin or its metabolites present in its milk (Yoshizawa et al. 1981). This value corresponded to a T-2 toxin equivalent of 11 ppb in milk, a level which probably would not be detrimental to human health. Longer term administration of large doses of T-2 toxin (equivalent to 50 ppm in animal feed) to a lactating cow (Robison et al. 1979b) resulted in levels of 10-80 ppb of T-2 toxin in the milk. Levels of other potentially toxic derivatives of T-2 toxin were not determined.

The transmission of radioactivity into eggs from hens administered single (0.25 mg/kg) or multiple (0.1 mg/kg) doses of [^{3}H]-T-2 toxin by oral intubation has also been studied (Chi et al. 1978b). The radioactivity in the egg reached a maximum about 24 h following a single dose and accounted for about 0.17% of the total administered radioactivity. The egg white contained about twice the specific activity as the yolk. For multiple dosed birds, up to 0.6% of the total administered radioactivity appeared in the eggs, the level reaching a plateau after 5 days of administration of the toxin. The amount of toxin in terms of T-2 equivalents per egg amounted to about 1 µg maximum, a level which probably presents no health hazzard to humans.

V. Metabolism

In vitro studies on the metabolic transformation of trichothecenes by liver enzymes have been reported. Ellison and Kotsonis (1974) reported that T-2 toxin could be specifically deacylated to HT-2 toxin by a microsomal enzyme fraction from liver. Ohta et al. (1977) demonstrated the specificity of the reaction and also showed that the enzyme could be found in microsomal fractions of kidney and spleen as well as liver from mice, rats, guinea-pigs and rabbits. The enzymatic activity was inhibited by eserine and diisopropyl-fluorophosphate implicating a non-specific carboxyesterase of microsomal origin in the hydrolysis (Ohta et al. 1977). Further studies on the substrate specificity of the microsomal esterase using 13 different trichothecenes was reported by Ohta et al. in 1978. The C-4 acetyl residues of diacetoxyscirpenol, T-2 toxin, fusaranon-X, and diacetylnivalenol were selectively hydrolyzed to the corresponding C-4 desacetyl derivatives. The C-3 acetyl group of monoacetyldeoxynivalenol and the C-8 acetyl group of 3,8-diacetylneosolaniol were also deacylated. Further hydrolysis of neosolaniol, HT-2 toxin, acetyl T-2 toxin and tetraacetylnivalenol could not be detected by the methods used; thus, these authors concluded that substitution at the C-3 or C-8 position of the trichothecene ring affects the selective enzymatic hydrolysis of acetyl groups on the C-4 position. The esterase from rabbit liver possessed a higher affinity than the rat liver enzyme for the class A trichothecenes while the rat liver enzyme possessed a higher affinity for the class B (8-keto) trichothecenes (Ohta et al. 1978).

Further studies on the metabolism of T-2 toxin by rat liver microsomal enzymes were performed using radiolabeled T-2 toxin (Yoshizawa et al. 1980c). T-2 toxin was hydrolyzed preferentially at the C-4 position to give HT-2 toxin which was further hydrolyzed to T-2 tetraol via a 4-desacetylneosolaniol intermediate. After one hour of incubation of the liver extract with T-2 toxin, the metabolite distribution was as follows: HT-2 toxin (C-4 desacetyl) (49.3%); C-4 desacetyl, C-8 desiosvaleryl (18.7%); T-2 tetraol (4.3%). The column methods used by these authors provided a more efficient method for obtaining the polar metabolites than the solvent partition methods used by Ohta et al. (1978). Rat intestinal strips were capable of the same transformations, but in addition a small amount of neosolaniol was formed from T-2 toxin indicating that hydrolysis of the isovaleryl group at C-8 might occur at a faster rate in the intestine than in the liver.

In vivo metabolic studies have been performed for T-2 toxin administered to chickens (Yoshizawa et al. 1980b) and a lactating cow (Yoshizawa et al. 1981). While small quantities of T-2 toxin and its desacylated derivatives (HT-2 toxin, neosolaniol and T-2 tetraol) were detected in excreta of chickens and in the plasma, milk, urine and feces of the cow, these metabolites were minor in comparison to three other metabolite peaks designated as TC-1, TC-3 and TC-6 (Yoshizawa et al. 1981). In at least two of these metabolites the epoxide ring is intact as demonstrated by a positive color reaction with 4-(p-nitrobenzyl) pyridine/tetraethylenepentamine reagent (Takitani et al. 1979). Thus, while some of the more polar metabolites (i.e., TC-6) may be formed in the liver via the glutathione-S-epoxidetransferase reaction as proposed by Foster et al. (1975), alternative conjugation systems which promote excretion but which keep the epoxide ring system intact must also be operating.

J. Trichothecenes – Pharmaceuticals or Chemical Weapons

I. Trichothecenes as Therapeutic Agents

Much of the recent research on the synthesis and chemical modifications of trichothecenes has as its goal the synthesis of derivatives with antineoplastic activity and relatively low animal toxicity (Still and Ohmizu 1981; Notegen et al. 1981; Tulshian and Fraser-Reid 1981; Banks et al. 1981). In a recent article, Doyle and Bradner (1980) reviewed the effects of many trichothecenes on the development of different leukemias in mice. Macrocyclic derivatives of the verrucarins and roridins, as well as many class A and class B trichothecenes, were tested. Recently, Jarvis et al. (1980c) tested many derivatives of the verrucarins and roridins, as well as the plant modified versions, the baccharins, for activity against the P-388 mouse leukemia. Most of the baccharinoids have a high T/C ratio – (days test animals lived/days control animals lived) x 100 – due apparently to the oxygen substituent on ring A. Jarvis et al. (1980c) introduced a β-9,10-epoxide into the A ring of verrucarin A and roridin A and substantially increased the T/C ratio for these compounds. Thus, continued modification and testing of the trichothecenes might result in development of a useful chemotherapeutic agent.

The potent effects of the trichothecenes on inhibition of protein synthesis coupled with the penetration powers of these compounds when applied topically to animals make them ideally suited for the treatment of topical viral infections. In the removal of warts, for example, one would like an agent which will rapidly inhibit viral replication, kill the tissue in the region of the infection, and have no long lasting effects on normal tissue regeneration beneath the dead tissue. Trichothecenes fulfill all of these requirements. Although there has been no development or testing of the trichothecenes for this use, they offer a valuable alternative to the podophyllin resins which are currently used for this purpose.

II. Trichothecenes as Chemical Weapons

The LD_{50} for T-2 toxin administered orally to rats is 3.8 mg/kg body wt. as compared to an LD_{50} of less than 8 mg/kg when applied in an organic solvent to the skin (Bamburg 1969). The closeness of these values indicates that the toxin is very effectively absorbed through the skin. This rapid absorption coupled with the low amounts of the toxin necessary to cause mental confusion and hallucinations in humans (about 100 µg/kg body wt.) makes these compounds potentially useful agents in chemical warfare. Although it is doubtful that rapid death would result from topical exposure or ingestion of trichothecenes, it is very possible that inhalation of these compounds in the form of an aerosol could result in pulmonary edema and/or hemorrhaging which may lead to death within a few hours. In addition, the nutritional state of the exposed individual may play a major role in his degree of susceptibility to the effects of the toxins. Highly sophisticated analytical methods have identified trichothecenes in samples of "yellow rain" and evidence has accumulated in support of the position that these compounds are not naturally occurring in the amounts and locations in which they have been found. Although the symptoms experienced by victims of these "yellow rain" attacks might be explained by the presence of trichothecenes in the samples, the high lethality reported might also indicate that other toxic agents are present, and their contribution to the symptoms and lethal nature of the attacks can not be ruled out.

Note Added in Proof: The first published reports of the analysis of "yellow-rain" samples have recently appeared (R.T. Rosen and J.D. Rosen, Biomed. Mass Spect. 9, 443 - 450 (1982). While providing definitive evidence on the presence of trichothecenes in the samples provided to the investigators, these reports are not in themselves indicative as to whether the trichothecenes were applied by man or developed through natural mechanisms. The presence of polyethylene glycol, a man made chemical, in the sample analyzed by Rosen and Rosen would suggest an unnatural origin of the sample if contamination with such an agent in the post collection procedures could be ruled out. Unfortunately, vials with rubber septa were used in the transport of this specimen and polyethylene glycol is used extensively in the production of rubber products. Further analysis of other samples will be necessary to answer this important question.

Two reviews of note have appeared since the submission of this manuscript: Jarvis, B.B., Mazzola, E.P.: Macrocyclic and other novel trichothecenes, their structure, synthesis and biological significance. Acc. Chem. Res. 15, 388-395 (1982); Ueno, Y. (Ed.): Trichothecenes-chemical, biological and toxicological aspects. Developments in Food Sci. 4, Amsterdam: Elsevier 1983.

References

Abrahamsson, S., Nilsson, B.: Direct determination of molecular structure of trichodermin. Proc. Chem. Soc., 188-189 (1964)

Abrahamsson, S., Nilsson, B.: The molecular structure of trichodermin. Acta Chem. Scand. 20, 1044-1052 (1966)

Achilladelis, B., Hanson, J.R.: Minor terpenoids of *Trichothecium roseum*. Phytochemistry 8, 765-767 (1969)

Agrelo, C.E., Schoental, R.: Synthesis of DNA in human fibroblasts treated with T-2 toxin and HT-2 toxin (the trichothecene metabolites of Fusarium species) and the effects of hydroxyurea. Toxicol. Lett. 5, 155-160 (1980)

Ames, B.N., Durston, W.E., Yamasaki, E., Lee, F.D.: Carcinogens are mutagens; a simple test system combining liver homogenates for activating and bacteria for detection. Proc. Natl. Acad. Sci. USA 70, 2281-2285 (1973)

Andrews, R.I., Thompson, B.K., Trenholm, H.L.: A national survey of mycotoxins in Canada. J. Am. Oil Chem. Soc. 58, 989A-991A (1981)

Bamburg, J.R.: Mycotoxins of the trichothecane family produced by cereal molds. Ph.D. Thesis, Univ. of Wisconsin, Madison (1969)

Bamburg, J.R.: Biological activity and detection of naturally occurring 12, 13-epoxy-Δ^9-trichothecenes. Clin. Toxicol. 5, 495-515 (1972)

Bamburg, J.R.: Chemical and biochemical studies of the trichothecene mycotoxins. Adv. Chem. 149, 144-162 (1976)

Bamburg, J.R., Strong, F.M.: Mycotoxins of the trichothecane family produced by *Fusarium tricinctum* and *Trichoderma lignorum*. Phytochemistry 8, 2405-2410 (1969)

Bamburg, J.R., Strong, F.M.: 12, 13-Epoxytrichothecenes. In: Microbial Toxins (eds. S. Kadis, A. Ciegler, S.J. Ajl), Vol. VII, pp. 207-292. New York: Academic Press 1971

Bamburg, J.R., Riggs, N.V., Strong, F.M.: The structures of toxins from two strains of *Fusarium tricinctum*. Tetrahedron 24, 3329-3336 (1968a)

Bamburg, J.R., Marasas, W.F.O., Riggs, N.V., Smalley, E.B., Strong, F.M.: Toxic spiroepoxy compounds from *Fusaria* and other Hyphomycetes. Biotechnol. Bioeng. 10, 445-455 (1968b)

Bamburg, J.R., Strong, F.M., Smalley, E.B.: Toxins from moldy cereals. J. Agric. Food Chem. 17, 443-450 (1969)

Banks, R.E., Miller, J.A., Nunn, M.J., Stanley, P., Weakley, J.R., Ullah, Z.: Diels-Alder route to potential trichothecene precursors. J. Chem. Soc. Perkin I, 1096-1102 (1981)

Barbacid, M., Vazquez, D.: Binding of [acetyl - ^{14}C] trichodermin to the peptidyl transferase center of eukaryotic ribosomes. Eur. J. Biochem. 44, 437-444 (1974)

Bawden, F.C., Freeman, G.G.: The natural behavior of inhibitors of plant viruses produced by *Trichothecium roseum* Link. J. Gen. Microbiol. 7, 154-160 (1952)

Beasley, V.R., Buck, W.B., Swanson, S.P., Szabo, J.R., Coppock, R.W., Burmeister, H.R., Vesonder, R.B.: Toxicokinetics of T-2 toxin in swine and cattle. J. Am. Vet. Med. Assoc. 179, 262 (1981)

Bennett, G.A., Peterson, R.E., Plattner, R.D., Shotwell, O.L.: Isolation and purification of deoxynivalenol and a new trichothecene by high pressure liquid chromatography. J. Am. Oil Chem. Soc. 58, 1002A-1005A (1981)

Betina, V., Vankova, M.: Trichothecin — an antibiotic, morphogenic factor, mycotoxin and bitter substance of apples. Biologia (Bratislava) 32, 943-949 (1977)

Bilai, V.I.: Experimental morphogenesis in *Fusaria* and their classification. Mikrobiol. Zh. 32, 158-163 (1970)

Bilai, V., Pidoplisko, N.M.: Toxigenic microscopical fungi. Kiev: Nankova Dumka 1970

Blight, M.M., Grove, J.F.: New metabolic products of *Fusarium culmorum*: toxic trichothec-9-en-8-ones and 2-acetylquinazolin-4(3H)-one. J. Chem. Soc. 1691-1693 (1974)

Böhner, B., Tamm, Ch.: Verrucarins and Roridins XI — The constitution of roridin A (insecticide from *Myrothecium roridum*). Helv. Chim. Acta 49, 2527-2546 (1966a)

Böhner, B., Tamm, Ch.: Verrucarins and Roridins XII — The constitution of roridin D. (from *Myrothecium roridum*). Helv. Chim. Acta 49, 2547-2554 (1966b)

Böhner, B., Fetz, E., Härri, E., Sigg, H.P., Stoll, Ch., Tamm, Ch.: Verrucarins and Roridins VIII — Über die Isolierung von Verrucarin H, Verrucarin J, Roridin D und Roridin E aus *Myrothecium*-Arten. Helv. Chim. Acta 48, 1079-1087 (1965)

Booth, C.: The genus *Fusarium*. Surry, England: Commonwealth Mycological Inst., Kew 1971

Bottalico, A.: Production of T-2 toxin by Fusarium spp. from cereals in Italy. Phytopathol. Medit. 16, 147-148 (1977)

Breitenstein, W., Tamm, Ch.: Verrucarin K, the first natural trichothecene derivative lacking the 12, 13-epoxy group. Helv. Chim. Acta 60, 1522-1527 (1977)

Brian, P.W., Dawkins, A.W., Grove, J.F., Hemming, H.G., Lowe, D., Norris, G.L.F.: Phytotoxic compounds produced by *Fusarium equiseti*. J. Exp. Bot. 12, 1-20 (1961)

Bukowski, R., Vaughn, C., Bottomley, R., Chen, T.: Phase II study of anguidine in gastrointestinal malignancies: a Southwest oncology group study. Cancer Treat. Rep. 66, 381-383 (1982)

Burmeister, H.R.: T-2 Toxin production by *Fusarium tricinctum* on solid substrate. Appl. Microbiol. 21, 739-742 (1971)

Burmeister, H.R., Vesonder, R.F., Kwolek, W.F.: Mouse bioassay for *Fusarium* metabolites: rejection or acceptance when dissolved in drinking water. Appl. Environ. Microbiol. 39, 957-961 (1980)

Cannon, M., Jimenez, A., Vazquez, D.: Competition between trichodermin and several other sesquiterpene antibiotics for binding to their receptor site(s) on eukaryotic ribosomes. Biochem. J. 160, 137-145 (1976)

Cannon, M., Cranston, W.I., Hellon, R.F., Townsend, Y.: Inhibition, by trichothecene antibiotics, of brain protein synthesis and fever in rabbits. J. Physiol. 322, 447-455 (1982)

Carrasco, L., Barbacid, M., Vazquez, D.: The Trichodermin group of antibiotics, inhibitors of peptide bond formation by eukaryotic ribosomes. Biochim. Biophys. Acta 312, 368-376 (1973)

Carter, C.J., Cannon, M.: Structure requirement for the inhibitory action of 12,13-epoxytrichothecenes on protein synthesis in eukaryotes. Biochem. J. 166, 399-409 (1977)

Carter, C.J., Cannon, M.: Inhibition of eukaryotic ribosomal function by the sesquiterpenoid antibiotic fusarenon-X. Eur. J. Biochem. 84, 103-111 (1978)

Chaney, S.G.: Principles of nutrition II. In: Textbook of Biochemistry with Clinical Correlations (ed. T.M. Devlin), pp. 1197-1239. New York: Wiley 1982

Chaytor, J.P., Saxby, M.J.: Development of a method for the analysis of T-2 toxin in maize by gas chromatography — mass spectroscopy. J. Chromatogr. 237, 107-113 (1982)

Chi., M.S., Mirocha, C.J., Kurtz, H.J., Weaver, G., Bates, F., Shimoda, W.: Effects of T-2 toxin on reproductive performance and health of laying hens. Poultry Sci. 56, 628-637 (1977a)

Chi, M.S., Mirocha, C.J., Kurtz, H.J., Weaver, G., Bates, F., Shimoda, W.: Subacute toxicity of T-2 toxin in broiler chicks. Poultry Sci. 56, 303-313 (1977b)

Chi, M.S., Robison, T.S., Mirocha, C.J., Swanson, S.P., Shimoda, W.: Excretion and tissue distribution of radioactivity from tritium-labeled T-2 toxin in chicks. Toxicol. Appl. Pharmacol. 45, 391-402 (1978a)

Chi, M.S., Robison, T.S., Mirocha, C.J., Behrens, J.C., Shimoda, W.: Transmission of radioactivity into eggs from laying hens (*Gallus domesticus*) administered tritium-labeled T-2 toxin. Poultry Sci. 57, 1234-1238 (1978b)

Chi, M.S., El-Halawani, M.E., Waibel, P.E., Mirocha, C.J.: Effects of T-2 toxin on brain catecholamines and selected blood components in growing chickens. Poultry Sci. 60, 137-141 (1981)

Chu, F.S., Grossman, S., Wei, R.D., Mirocha, C.J.: Production of antibody against T-2 toxin. Appl. Environ. Microbiol. 37, 104-108 (1979)

Chung, C.W., Trucksess, M.W., Giles, A.L., Jr., Friedman, L.: Rabbit skin test for estimation of T-2 toxin and other skin-irritating toxins in contaminated corn. J. Assoc. Off. Anal. Chem. 57, 1121-1127 (1974)

Ciegler, A.: Mycotoxins — their biosynthesis in fungi: biosynthesis of the trichothecenes. J. Food Protect. 42, 825-828 (1979)

Claridge, C.A., Schmitz, H.: Microbial and chemical transformations of some 12,13-epoxytrichothecene-9, 10-enes. Appl. Environ. Microbiol. 36, 63-67 (1978)

Claridge, C.A., Schmitz, H.: Production of 3-acetoxyscirpene -4,15-diol from anguidine (4,15-diacetoxyscirpene-3-ol) by *Fusarium oxysporum* F. sp. vasinfectum. Appl. Environ. Microbiol. 37, 693-696 (1979)

Cole, R.J., Dorner, J.W., Cox, R.H., Cunfer, B.M., Cutler, H.G., Stuart, B.P.: The isolation and identification of several trichothecene mycotoxins from *Fusarium heterosporum*. J. Nat. Prod. (Lloydia) 44, 324-330 (1981)

Collins, G.J., Rosen, J.D.: Distribution of T-2 toxin in wet-milled corn products. J. Food Sci. 46, 877-879 (1981)

Colvin, E.W., Raphael, R.A., Roberts, J.S.: The total synthesis of (±)-trichodermin. Chem. Commun. 858-859 (1971)
Colvin, E.W., Malchenko, S., Raphael, R.A., Roberts, J.S.: Synthetic studies on the sesquiterpene antibiotic verrucarol. J. Chem. Soc. Perkin Trans. 1, 658-662 (1978)
Coss, R.A., Bamburg, J.R., Dewey, W.C.: The effects of X irradiation on microtubule assembly in vitro. Radiat. Res. 85, 99-115 (1981)
Cullen, D., Smalley, E.B.: New process for T-2 toxin production. Phytopathol. 71, 212 (1981)
Cundliffe, E., Davies, J.: Inhibition of initiation, elongation and termination of eukaryotic protein synthesis by trichothecene fungal toxins. Antimicrob. Agents Chemother. 11, 491-499 (1977)
Cundliffe, E., Cannon, M., Davies, J.: Mechanisms of inhibition of eukaryotic protein synthesis by trichothecene fungal toxins. Proc. Natl. Acad. Sci. USA 71, 30-34 (1974)
Cutler, H.G., LeFiles, J.H.: Trichodermin: effects on plants. Plant Cell Physiol. 19, 177-182 (1978)
Davis, G.R.F., Westcott, N.D., Smith, J.D., Neish, G.A., Schiefer, H.B.: Toxigenic isolates of *Fusarium sporotrichioides* obtained from hay in Saskatchewan. Can. J. Microbiol. 28, 259-261 (1982)
Dawkins, A.W., Grove, J.F., Tidd, B.K.: Diacetoxyscirpenol and some related compounds. Chem. Commun. 27-28 (1965)
DeNicola, D.B., Rebar, A.H., Carlton, W.W., Yagen, B.: T-2 Toxin mycotoxicoses in the guinea-pig. Food Cosmet. Toxicol. 16, 601-609 (1978)
Doerr, J.A., Hamilton, P.B., Burmeister, H.R.: T-2 Toxicosis and blood coagulation in young chickens. Toxicol. Appl. Pharmacol. 60, 157-162 (1981)
Dosik, G., Barlogie, B., Johnston, D.A., Murphy, W.K., Drewinko, B.: Lethal and cytokinetic effects of anguidin on a human colon cancer cell line. Can. Res. 38, 3304-3309 (1978)
Dounin, M.: The Fusariosis of cereal crops in European Russia in 1923. Phytopathol. 16, 305-308 (1926)
Doyle, T.W., Bradner, W.T.: Trichothecanes. In: Anticancer Agents Based on Natural Product Models. Medicinal Chemistry (eds. J.M. Cassady, J.D. Douros), Vol. 16, pp. 43-72. New York: Academic Press 1980
Drobotka, V.G., Maruschenko, P.E., Aizeman, B.E., Kolesnik, N.G., Kudlai, D.B., Tatel, P.D., Mehnichenko, V.D.: Stachybotryotoxicosis, a new disease of horses and humans. Am. Rev. Sov. Med. 2, 238-242 (1945)
Ellison, R.A., Kotsonis, F.N.: In vitro metabolism of T-2 toxin. Appl. Microbiol. 27, 423-424 (1974)
Eppley, R.M.: Methods for the detection of trichothecenes. J. Assoc. Off. Anal. Chem. 58, 906-908 (1975)
Eppley, R.M.: Trichothecenes and their analysis. J. Am. Oil Chem. Soc. 56, 824-829 (1979)
Eppley, R.M., Bailey, W.J.: 12,13-Epoxy-Δ^9-trichothecenes as the probable mycotoxins responsible for stachybotryotoxicosis. Science 181, 758-760 (1973)
Eppley, R.M., Mazzola, E.P., Highet, R.J., Bailey, W.J.: Structure of Satratoxin H, a metabolite of *Stachybotrys atra*. Application of proton and carbon-13 nuclear magnetic resonance. J. Org. Chem. 42, 240-243 (1977)
Eppley, R.M., Mazzola, E.P., Stack, M.E., Dreifuss, P.A.: Structure of Satratoxin F and Satratoxin G, metabolites of *Stachybotrys atra*: Application of proton and carbon-13 nuclear magnetic resonance spectroscopy. J. Org. Chem. 45, 2522-2523 (1980)
Fetz, E., Böhner, B., Tamm, Ch.: Verrucarins and Roridins IX: Die Konstitution von Verrucarin J. Helv. Chim. Acta 48, 1669-1679 (1965)
Flury, E., Mauli, R., Sigg, H.P.: The constitution of diacetoxyscirpenol. Chem. Commun. 26-27 (1965)

Forgacs, J.: Stachybotryotoxicosis and moldy corn toxicosis. In: Mycotixins in Foodstuffs (ed. G.N. Wogan), pp. 87-104. Cambridge, MA: MIT Press 1965
Forgacs, J., Carll, W.T.: Mycotoxicoses. Adv. Vet. Sci. 7, 273-282 (1962)
Forgacs, J., Carll, W.T., Herring, A.S., Mahlandt, B.G.: Toxicity of *Stachybotrys atra* for animals. Trans. N.Y. Acad. Sci. 20, 787-808 (1958)
Foster, P.M.D., Slater, T.F., Patterson, D.S.P.: A possible enzymatic assay for trichothecene mycotoxins in animal feedstuffs. Biochem. Soc. Trans. 3, 875-878 (1975)
Freeman, G.G., Gill, J.E.: Alkaline hydrolysis of trichothecin. Nature 166, 698-699 (1950)
Freeman, G.G., Morrison, R.I.: Trichothecin: An antifungal metabolic product of *T. roseum* Link. Nature 162, 30 (1948)
Freeman, G.G., Morrison, R.I.: The isolation and chemical properties of trichothecin, an antifungal substance from *Trichothecium roseum* Link. Biochem. J. 44, 1-5 (1949)
Fried, H.M., Warner, J.R.: Cloning of yeast gene for trichodermin resistance and ribosomal protein L3. Proc. Natl. Acad. Sci. USA 78, 238-242 (1981)
Fromentin, H., Salazar-Mejicanos, S., Mariat, F.: Pouvoir pathogêne de *Candida albicans* pour la souris normale ou dêprimée par une mycotoxine: le diacetoxyscirpenol. Ann. Microbiol. 131B, 39-46 (1980)
Fromentin, H., Salazar-Mejicanos, S., Mariat, F.: Experimental cryptococcosis in mice treated with diacetoxyscirpenol, a mycotoxin of *Fusarium*. Sabouraudia 19, 311-313 (1981)
Gardner, D., Glen, A.T., Turner, W.B.: Calonectrin and 15-deacetyl-calonectrin, new trichothecanes from *Calonectria nivalis*. J. Chem. Soc. Perkin Trans. I, 2576-2578 (1972)
Gentry, P.A., Cooper, M.L.: Effect of Fusarium T-2 toxin on hematological and biochemical parameters in the rabbit. Can. J. Comp. Med. 45, 400-405 (1981)
Ghosal, S., Chakrabarti, D.K., Basu Chaudhary, K.C.: Toxic substances produced by *Fusarium I*: Trichothecene derivatives from two strains of *Fusarium oxysporum* f. sp. carthami. J. Pharm. Sci. 65, 160-161 (1976)
Ghosal, S., Biswas, K., Srivastava, R.S., Chakrabarti, D.K., Basu Chaudhary, K.C.: Toxic substances produced by Fusarium V. Occurance of zearalenone, diacetoxyscirpenol and T-2 toxin in moldy corn infected with *Fusarium moniliforme* Sheld. J. Pharm. Sci. 67, 1768-1769 (1978)
Ghosal, S., Chakrabarti, D.K., Srivastava, A.K., Srivastava, R.S.: Toxic 12,13-epoxytrichothecenes from anise fruits infected with *Trichothecium roseum*. J. Agric. Food Chem. 30, 106-109 (1982)
Gláz, E.T., Csanyi, E., Gyimesi, J.: Supplementary data on crotocin, an antifungal antibiotic. Nature 212, 617-618 (1966)
Godtfredsen, W.O., Vangedal, S.: Trichodermin, a new antibiotic related to trichothecin. Proc. Chem. Soc. 188-189 (1964)
Godtfredsen, W.O., Vangedal, S.: Trichodermin, a new sesquiterpene antibiotic. Acta Chem. Scand. 19, 1088-1102 (1965)
Godtfredsen, W.O., Grove, J.F., Tamm, Ch.: On the nomenclature of a new class of sesquiterpenes. Helv. Chim. Acta 50, 1666-1668 (1967)
Goodwin, W., Haas, C.D., Fabian, C., Heller-Bettinger, I., Hoogstraten, B.: Phase I evaluation of anguidine (diacetoxyscirpenol, NSC-141537). Cancer 42, 23-26 (1978)
Grant, P.G., Schindler, D., Davies, J.E.: Mapping of trichodermin resistance in *Saccharomyces cerevisiae*: A genetic locus for a component of the 60s ribosomal subunit. Genetics 83, 667-673 (1976)
Greenway, J.A., Puls, R.: Fusaritoxicosis from barley in British Columbia I. Natural occurrence and diagnosis. Can. J. Comp. Med. 40, 12-15 (1976)

Grove, J.F.: Phytotoxic compounds produced by *Fusarium equiseti*. Part VI. 4-beta-8-alpha, 15-triacetoxy-12,13-epoxytrichothec-9-ene-3-alpha, 7-alpha-diol. J. Chem. Soc. Sect. C Org. Chem. 2, 378-379 (1970a)

Grove, J.F.: Phytotoxic compounds produced by *Fusarium euqiseti*. Part V. Transformation products of 4β-15-diacetoxy-3α-7α-dihydro-12,13-epoxytrichothec-9-ene-8-one and the structure of nivalenol and fusarenone. J. Chem. Soc. Sect. C Org. Chem. 2, 375-378 (1970b)

Grove, J.F., Hoskin, M.: The larvicidal activity of some 12,13-epoxy-trichothec-9-enes. Biochem. Pharmacol. 24, 959-962 (1975)

Grove, J.F., Mortimer, P.H.: The cytotoxicity of some transformation products of diacetoxyscirpenol. Biochem. Pharmacol. 18, 1473-1478 (1969)

Gutzwiller, J., Tamm, Ch.: Verrucarins and Roridins V: Über die Struktur von Verrucarin A. Helv. Chim. Acta 48, 157-176 (1965a)

Gutzwiller, J., Tamm, Ch.: Verrucarins and Roridins VI: Über die Struktur von Verrucarin B. Helv. Chim. Acta 48, 177-182 (1965b)

Gutzwiller, J., Mauli, R., Sigg, H.P., Tamm, Ch.: Die Konstitution von Verrucarol und Roridin C. Helv. Chim. Acta 47, 2243-2262 (1964)

Gyimesi, J., Melera, A.: On the structure of crotocin, an antifungal antibiotic. Tetrahedron Lett. 1665-1673 (1967)

Harrach, B., Mirocha, C.J., Pathre, S.V., Palyusik, M.: Macrocyclic trichothecene toxins produced by a strain of *Stachybotrys atra* from Hungary. Appl. Environ. Micobiol. 41, 1428-1432 (1981)

Härri, E., Loeffler, W., Sigg, H.P., Stähelin, H., Stoll, Ch., Tamm, Ch., Wiesinger, D.: Über die Verrucarine und Roridine, eine Gruppe von cytostatisch hochwirksamen Antibiotica aus *Myrothecium*-Arten. Helv. Chim. Acta 45, 839-853 (1962)

Hayes, A.W., Hood, R.D., Lee, H.L.: Teratogenic effects of ochratoxin A in mice. Teratology 9, 93-98 (1974)

Hayes, M.A., Schiefer, H.B.: Quantitative and morphological aspects of cutaneous irritation by trichothecene mycotoxins. Food Cosmet. Toxicol. 17, 611-621 (1979)

Hayes, M.A., Schiefer, H.B.: Subacute toxicity of dietary T-2 toxin in mice: Influence of protein nutrition. Can. J. Comp. Med. 44, 219-228 (1980)

Hayes, M.A., Bellamy, J.E.C., Schiefer, H.B.: Subacute toxicity of dietary T-2 toxin in mice: Morphological and hematological effects. Can. J. Comp. Med. 44, 203-218 (1980)

Hintikka, E.-L.: Stachybotryotoxicosis as a veterinary problem. In: Mycotoxins in Human and Animal Health (eds. J.V. Rodricks, C.W. Hesseltine, M.A. Mehlman). Park Forest South. IL: Pathotox Publishers 1977

Hood, R.D., Kuczuk, M.H., Szczech, G.M.: Effects in mice of stimultaneous prenatal exposure to ochratoxin A and T-2 toxin. Teratology 17, 25-30 (1978)

Hsu, I.-C., Smalley, E.B., Strong, F.M., Ribelin, W.E.: Identification of T-2 toxin in moldy corn associated with a lethal toxicoses in dairy cattle. Appl. Microbiol. 24, 684-690 (1972)

Huff, W.E., Doerr, J.A., Hamilton, P.B., Vesonder, R.F.: Acute toxicity of vomitoxin (deoxynivalenol) in broiler chickens. Poultry Sci. 60, 1412-1414 (1981)

Ilus, T., Ward, P.J., Nummi, M., Adlercreutz, H., Gripenberg, J.: A new mycotoxin from *Fusarium*. Phytochemistry 16, 1839-1840 (1977)

Ishii, K.: Two new trichothecenes produced by *Fusarium* sp. Phytochemistry 14, 2469-2471 (1975)

Ishii, K., Ueno, Y.: Isolation and characterization of two new trichothecenes from *Fusarium sporotrichioides* strain M-1-1. Appl. Environ. Microbiol. 42, 541-543 (1981)

Ishii, K., Ando, Y., Ueno, Y.: Toxicological approaches to the metabolites of *Fusaria*. IX. Isolation of vomiting factor from moldy corn infected with *Fusarium* species. Chem. Pharm. Bull. 23, 2162-2164 (1975)

Ishii, K., Pathre, S.V., Mirocha, C.J.: Two new trichothecenes produced by *Fusarium roseum*. J. Agric. Food Chem. 26, 649-653 (1978)
Ito, T., Tashiro, F., Kajiwara, S., Ueno, Y.: Bindings of T-2 toxin and other mycotoxins with human serum protein. Toxicon 17, Suppl. 1, 76 (1979)
Jagadeesan, V., Rukmini, C., Vijayaraghavan, M., Tulpule, P.G.: Immune studies with T-2 toxin: effect of feeding and withdrawal in monkeys. Food Chem. Toxicol. 20, 83-87 (1982)
Jarvis, B.B., Stahly, G.P., Pavanasasivam, G., Mazzola, E.P.: Structure of roridin J, a new macrocyclic trichothecene from *Myrothecium verrucaria*. J. Antibiot. (Tokyo) 33, 256-258 (1980a)
Jarvis, B.B., Midiwo, J., Stahly, G.P., Pavanasasivam, G., Mazzola, E.P.: Trichodermadiene: a new trichothecene. Tetrahedron Lett. 21, 787-788 (1980b)
Jarvis, B.B., Stahly, G.P., Pavanasasivam, G., Mazzola, E.P.: Antileukemic compounds derived from the chemical modification of macrocyclic trichothecenes. 1. Derivatives of verrucarin A. J. Med. Chem. 23, 1054-1058 (1980c)
Jarvis, B.B., Midiwo, J.O., DeSilva, T., Mazzola, E.P.: Verrucarin L., a new macrocyclic trichothecene. J. Antibiot. (Tokyo) 34, 120-121 (1981a)
Jarvis, B.B., Pavanasasivam, G., Holmlund, C.E., Desilva, T., Stahly, G.P., Mazzola, E.P.: Biosynthetic intermediates to the macrocyclic trichothecenes. J. Am. Chem. Soc. 103, 472-474 (1981b)
Jarvis, B.B., Midiwo, J.O., Tuthill, D., Bean, G.A.: Interaction between the antibiotic trichothecenes and the higher plant *Baccharis Megapotamica*. Science 214, 460-462 (1981c)
Jarvis, B.B., Midiwo, J.O., Flippen-Anderson, J.L., Mazzola, E.P.: Stereochemistry of the roridins. J. Natur. Prod. 45, 440-448 (1982)
Jemmali, M., Ueno, Y., Ishii, K., Frayssinet, C., Etienne, M.: Natural occurrence of trichothecenes (nivalenol, deoxynivalenol, T_2) and zearalenone in corn. Experientia 34, 1333-1334 (1978)
Joffe, A.Z.: Biological properties of some toxic fungi isolated from overwintered cereals. Mycopathol. Mycol. Appl. 16, 201-221 (1962)
Joffe, A.Z.: Toxin production by cereal fungi causing toxic alimentary aleukia in man. In: Mycotoxins in Foodstuffs (ed. G.N. Wogan), pp.77-85. Cambridge, MA: MIT Press 1965
Joffe, A.Z., Palti, J.: Taxonomic studiy of *Fusaria* of the Sporotrichiella section used in recent toxicological work. Appl. Microbiol. 29, 575-579 (1975)
Joffe, A.Z., Yagen, B.: Intoxication produced by toxic fungi *Fusarium poae* and *F. sporotrichioides* on chicks. Toxicon 16, 263-273 (1978)
Jones, E.R.H., Lowe, G.: The biogenesis of trichothecin. J. Chem. Soc. 3959-3962 (1960)
Kamimura, H., Nishijima, M., Yasuda, K., Saito, K., Ibe, A., Nagayama, T., Ushiyama, H., Naoi, Y.: Simultaneous detection of several Fusarium mycotoxins in cereals, grains, and foodstuffs. J. Assoc. Off. Anal. Chem. 64, 1067-1073 (1981)
Korpinen, E.-L.: Studies on *Stachybotrys alternans* IV. Effect of low doses of stachybotrys toxins on pregnancy of mice. Acta Pathol. Microbiol. Scand. Sect. B 82, 457-464 (1974)
Kosuri, N.R., Grove, M.D., Yates, S.G., Tallent, W.H., Ellis, J.J., Wolff, I.A., Nichols, R.E.: Response of cattle to mycotoxins of *Fusarium tricinctum* isolated from corn and fescue. J. Am. Vet. Med. Assoc. 157, 938-940 (1970)
Kosuri, N.R., Smalley, E.B., Nichols, R.E.: Toxicological studies of *Fusarium tricinctum* (Corda) Snyder and Hansen from moldy corn. Am. J. Vet. Res. 32, 1843-1850 (1971)
Kotsonis, F.N., Ellison, R.A.: Assay and Relationship of HT-2 toxin and T-2 toxin formation in liquid culture. Appl. Microbiol. 30, 33-37 (1975)
Kotsonis, F.N., Smalley, E.B., Ellison, R.H., Gale, C.M.: Feed refusal factors in pure cultures of *Fusarium roseum 'graminearum'*. Appl. Microbiol. 30, 362-368 (1975a)

Kotsonis, F.N., Ellison, R.H., Smalley, E.B.: Isolation of acetyl T-2 toxin from *Fusarium poae*. Appl. Microbiol. 30, 493-495 (1975b)

Kriegleder, H.: Morphological findings in guinea pigs after acute and subacute intoxication with diacetoxyscirpenol. Zentralbl. Veterinaermed. Reihe A 28, 165-175 (1981)

Kuczuk, M.H., Benson, P.M., Heath, H., Hayes, A.W.: Evaluation of the mutagenic potential of mycotoxins using *Salmonella typhimurium* and *Saccharomyces cerevisiae*. Mutat. Res. 53, 11-20 (1978)

Kupchan, S.M., Jarvis, B.B., Dailey, R.G., Bright, W., Bryan, R.F., Shizari, Y.: Baccharin, a novel potent antileukemic trichothecene triepoxide from *Baccharis megapotamica*. J. Am. Chem. Soc. 98, 7092-7093 (1976)

Kupchan, S.M., Streelman, D.R., Jarvis, B.B., Dailey, R.G., Jr., Sneden, A.T.: Isolation of potent new antileukemic trichothecenes from *Baccharis megapotamica*. J. Org. Chem. 12, 4221-4225 (1977)

Kuriyama, R., Sakai, H.: Role of tubulin - SH groups in polymerization to microtubules. J. Biochem. (Tokyo) 76, 651-654 (1974)

Kurtz, H.J., Mirocha, C.J., Meade, R.: Proc. 4th International Congress Pig Vet. Soc., Ames IA, p. N6, June 22-24, 1976

Lafarge-Frayssinet, C., Lespinats, G., Lafont, P., Loisillier, F., Mousset, S., Rosenstein, Y., Frayssinet, C.: Immunosuppressive effects of *Fusarium* extracts and trichothecenes: Blastogenic response of murine splenic and thymic cells to mitogens. Proc. Soc. Exp. Biol. Med. 106, 302-311 (1979)

Lafarge-Frayssinet, C., Chakor, K., Lafont, P., Frayssinet, C.: Transplacental transfer of T-2 toxin: Pathological effect. In: Proc. 4th Int. IUPAC Symp. on Mycotoxins and Phycotoxins (eds. P. Krogh, G.H. Scherr). Park Forest South, IL: Pathotox Publishers 1980

Lafarge-Frayssinet, C., Decloite, F., Mousset, S., Martin, M., Frayssinet, C.: Induction of DNA single-strand breaks by T_2 toxin, a trichothecene metabolite of *Fusarium*. Effect on lymphoid organs and liver. Mutation Res. 88, 115-123 (1981)

Lansden, J.A., Cole, R.J., Dorner, J.W., Cox, R.H., Cutler, H.G., Clark, J.D.: A new trichothecene mycotoxin isolated from *Fusarium tricinctum*. J. Agric. Food Chem. 26, 246-249 (1978)

Lee, S., Chu, F.S.: Radioimmunoassay of T-2 toxin in biological fluids. J. Assoc. Off. Anal. Chem. 64, 684-688 (1981a)

Lee, S., Chu, F.S.: Radioimmunoassay of T-2 toxin in corn and wheat. J. Assoc. Off. Anal. Chem. 64, 156-161 (1981b)

Lindenfelser, L.A., Lillehoj, E.B., Burmeister, H.R.: Aflatoxin and tricothecene toxins; skin tumor induction and synergistic acute toxicity in white mice. J. Natl. Cancer Inst. 52, 113-116 (1974)

Linnainmaa, K., Sorsa, M., Ilus, T.: Epoxytrichothecene mycotoxins as C-mitotic agents in *Allium*. Hereditas 90, 151-156 (1979)

Lo, A.C., McDougall, J., Nazar, R.N.: Altered ribosomal proteins in trichodermin (Resistant) strain of yeast. Can. J. Genet. 22, 670 (1980)

Löffler, W., Mauli, R., Rüsch, M.E., Stähelin, H.: Anguidin and derivatives, a new antibiotic and antitumoral product. Chem. Abst. 62, 5856d (1965)

Lutsky, I., Mor, N., Yagen, B., Joffe, A.Z.: The role of T-2 toxin in experimental alimentary toxic aleukia: A toxicity study in cats. Toxicol. Appl. Pharmacol. 43, 111-124 (1978)

Machida, Y., Nozoe, S.: Biosynthesis of trichothecin and related compounds. Tetrahedron 28, 5113-5117 (1972)

Mains, E.B., Vestal, C.M., Curtis, P.B.: Scab of small grains and feeding trouble in Indiana in 1928. Indiana Acad. Sci. Proc. 39, 101-110 (1930)

Mann, D.D., Buening, G.M., Hook, B.S., Osweiler, G.D.: Effect of T-2 toxin on the bovine immune system: Humoral factors. Infect. Immun. 36, 1249-1252 (1982)

Marasas, W.F.O.: Moldy corn: Nutritive value, toxicity and mycoflora with reference to *Fusarium tricinctum* (Corda) Snyder et Hansen. Ph.D. Thesis, University of Wisconsin, MA 1969

Marasas, W.F.O., Bamburg, J.R., Smalley, E.B., Strong, F.M., Ragland, W.L., Degurse, P.E.: The effects on trout, rats and mice of T-2 toxin produced by the fungus *Fusarium tricinctum* (Cd.) Snyder et Hansen. Toxicol. Appl. Pharmacol. 15, 471-482 (1969)

Marasas, W.F.O., Smalley, E.B., Bamburg, J.R., Strong, F.M.: Phytotoxicity of T-2 toxin produced by *Fusarium tricinctum*. Phytopathology 61, 1488-1491 (1971)

Marasas, W.F.O., Kriek, N.P., Steyn, M., van Rensburg, S.J., van Schalkwyk, D.J.: Mycotoxicological investigation on Zambian maize. Food Cosmet. Toxicol. 16, 39-45 (1978)

Marasas, W.F.O., Leistner, L., Hofmann, G., Eckard, C.: Occurrence of toxigenic strains of *Fusarium* in maize and barley in Germany. Eur. J. Appl. Microbiol. Biotechnol. 7, 289-305 (1979a)

Marasas, W.F.O., van Rensburg, S.J., Mirocha, C.J.: Incidence of *Fusarium* species and the mycotoxins deoxynvalenol and zearalenene in corn produced in esophageal cancer areas in Transkei. J. Agric. Food Chem. 27, 1108-1112 (1979b)

Masuda, E., Takemoto, T., Tatsuno, T., Obara, T.: Immunosuppressive effect of a trichothecene mycotoxin, fusarenon-X in mice. Immunology 45, 743-749 (1982)

Masuko, H., Ueno, Y., Otokawa, M., Yaginuma, K.: The enhancing effect of T-2 toxin on delayed hypersensitivity in mice. Jap. J. Med. Sci. Biol. 30, 159-163 (1977)

Matsumoto, M., Minato, H., Tori, K., Ueyama, M.: Structures of isororidin E and epoxy and diepoxyroridin H, new metabolites isolated from *Cylindrocarpon* species determined by cabon-13 and hydrogen-1 NMR spectroscopy. Revision of the C-2', C-3' double bond configuration in the roridin group. Tetrahedron Lett. 4093-4096 (1977)

Matsumoto, H., Ito, T., Ueno, Y.: Toxicological approaches to the metabolite of *Fusaria* XII. Fate and distribution of T-2 toxin in mice. Jap. J. Exp. Med. 48, 393-399 (1978)

Matsuoka, Y., Kubota, K.: Studies on mechanisms of diarrhea induced by fusarenon-X, a trichothecene mycotoxin from *Fusarium* species. Toxicol. Appl. Pharmacol. 57, 293-301 (1981)

Matsuoka, Y., Kubota, K.: Studies on mechanisms of diarrhea induced by fusarenon-X, a trichothecene mycotoxin from *Fusarium* species; characteristics of increased intestinal absorption rate induced by fusarenon-X. J. Pharmacobio-Dyn. 5, 193-199 (1982)

Matsuoka, Y., Kubota, K., Ueno, Y.: General pharmacological studies of fusarenon-X, a trichothecene mycotoxin from *Fusarium* species. Toxicol. Appl. Pharmacol. 50, 87-94 (1979)

Matthews, J.G., Patterson, D.S.P., Roberts, B.A., Shreeve, B.J.: T-2 toxin and haemorrhagic syndromes of cattle. Vet. Record 101, 391 (1977)

Matthews, J.G., Shreeve, B.J., Patterson, D.S.P., Hayes, A.W.: Experimental studies on haemorrhagic disease. Proc. 3rd Mtg. Mycotoxins in Animal Dis., pp. 54-57. Weybridge: Ministery of Agriculture, Fisheries, and Food; Agricultural Service 1979

Mayer, C.F.: Endemic panmyelotoxicosis in the Russian grain belt. Part I. The clinical aspects of alimentary toxic aleukia (ATA); a comprehensive review. Military Surg. 113, 173-189 (1953)

McLaughlin, C.S., Vaughan, M.H., Campbell, I.M., Wei, C.M., Stafford, M.E., Hansen, B.S.: Inhibition of protein synthesis by trichothecenes. In: Mycotoxins in Human and Animal Health (eds. J.V. Rodricks, C.W. Hesseltine, M.A. Mehlman), pp. 263-273. Park Forest South, IL: Pathotox Publishers 1977

McLean, A.E.M., McLean, E.K.: Diet and toxicity. Br. Med. Bull. 25, 278-281 (1969)
McPhail, A.T., Sim, G.A.: Verrucarin A — absolute configuration. J. Chem. Soc. Sect. C Org. Chem. 1394-1406 (1966)
Mellon, M., Rebhun, L.I.: Studies on the accessible sulfhydryls of polymerizable tubulin. In: Cell Motility (eds. R. Goldman, T. Pollard, J. Rosenbaum), Book C, Microtubules and Related Proteins, pp. 1149-1163. New York: Cold Spring Harbor Laboratory 1976
Midiwo, J.O.: Part I. Mechanism of the thermal decomposition of α-azido sulfoxides. Part II. Isolation, elucidation and chemical modifications of tricothecenes from *Myrothecium verrucaria*. Ph. D. Thesis, Univ. of Maryland, College Park, MD 1981
Minato, H., Katayama, T., Tori, K.: Vertisporin, a new antibiotic from *Verticimonosporium diffractum*. Tetrahedron Lett. 30, 2579-2582 (1975)
Mirocha, C.J., Christensen, C.M.: Oestrogenic mycotoxins synthesized by *Fusarium*. In: Mycotoxins (ed. I.F.H. Purchase), pp. 129-148. Amsterdam: Elsevier 1974
Mirocha, C.J., Pathre, S.V.: Identification of the toxic principle in a sample of poaefusarin. Appl. Microbiol. 26, 719-724 (1973)
Mirocha, C.J., Pathre, S.V., Schauerhamer, B., Christensen, C.M.: Natural occurrence of *Fusarium* toxins in feedstuff. Appl. Environ. Microbiol. 32, 553-556 (1976)
Mizuno, S.: Mechanism of inhibition of protein synthesis initiation by diacetoxyscirpenol and fusarenon-X in the reticulocyte lysate system. Biochim. Biophys. Acta 383, 207-214 (1975)
Morooka, N., Uratsyji, N., Yoshizawa, T., Yamamoto, H.: Studies on the toxic substances in barley infected with *Fusarium spp*. Jap. J. Food Hyg. 13, 368-375 (1972)
Nagao, M., Honda, M., Hamasaki, T., Natori, S., Ueno, Y., Yamasaki, M., Seino, Y., Yahagi, T., Sugimura, T.: Mutogenicity of mycotoxins on *Salmonella*. Proc. Jap. Assoc. Mycotoxicol. 3/4, 41-43 (1976)
Nakamura, Y., Ohta, M., Ueno, Y.: Reactivity of 12,13-epoxytrichothecenes with epoxide hydrolase, glutathione-S-transferase and glutathione. Chem. Pharm. Bull. 25, 3410-3414 (1977)
Norppa, H., Penttila, M., Sorsa, M., Hintikka, E.-L., Ilus, T.: Mycotoxin T-2 of *Fusarium tricinctum* and chromosome changes in Chinese hamster bone marrow. Hereditas 93, 329-332 (1980)
Notegen, E.-A., Tori, M., Tamm, C.: Partial synthesis of 3'-hydroxy-2'-deoxy-2", 3", 4", 5" - tetrahydroverrucarin A. Helv. Chim. Acta 64, 316-328 (1981)
Ohta, M., Ishii, K., Ueno, Y.: Metabolism of trichothecene mycotoxins I. Microsomal deacetylation of T-2 toxin in animal tissues. J. Biochem. (Tokyo) 82, 1591-1598 (1977)
Ohta, M., Matsumoto, H., Ishii, K., Ueno, Y.: Metabolism of trichothecene mycotoxins II. Substrate specificity of microsomal deacetylation of trichothecenes. J. Biochem. (Tokyo) 84, 697-706 (1978)
Ohtsubo, K., Yamada, M., Saito, M.: Inhibitory effects of nivalenol, a toxic principle of *Fusarium nivale*, on the growth and biopolymer synthesis of HeLa cells. Jap. J. Med. Sci. Bull. 21, 185-194 (1968)
Ohtsubo, K., Kaden, P., Mittermayer, C.: Polyribosomal breakdown in mouse fibroblasts (L-cells) by fusarenon-X, a toxic principle isolated from *Fusarium nivale*. Biochim. Biophys. Acta 287, 520-525 (1972)
Okuchi, M., Itoh, M., Kaneko, Y., Doi, S.: A new antifungal substance produced by *Myrothecium*. Agr. Biol. Chem. 32, 394-395 (1968)
Oldham, J.W., Allred, L.E., Milo, G.E., Kindig, O., Capen, C.C.: The toxicological evaluation of the mycotoxins T-2 and T-2 tetraol using normal human fibroblasts in vitro. Toxicol. Appl. Pharmacol. 52, 159-168 (1980)

Olifson, L.E.: Chemical action of some fungi on overwintered cereals. Monitor, Orenburg Sect. of the USSR D.J. Mendeleyev Chem. Soc. 7, 21-35 (1957)
Olifson, L.E., In: Mycotoxicoses of man and agricultural animals (ed. V.I. Bilai), p. 58. Kiev, USSR: Acad. Sci. UKr. SSR 1960
Otokawa, M., Shibahara, Y., Egarashi, Y.: The inhibitory effect of T-2 toxin on tolerance induction of delayed-type hypersensitivity in mice. Jap. J. Med. Sci. Biol. 32, 34-45 (1979)
Pääkkonen, R., Ronnholm, K., Raunio, V.: The nonmutagenicity of some mycotoxins by *Salmonella*/microsome assay. Acta Univ. Tamper. Ser. B 9, 50 (1978)
Pace, J.G., Murphy, P.E.: In vitro and in vivo effects of T-2 toxin on mitochondrial respiration. Fed. Proc. 41, 525 (1982)
Palmisano, F., Visconti, A., Bottalico, A., Lerario, P., Zambonin, P.G.: Differential-pulse polarography of trichothecene toxins: Detection of deoxynivalenol in corn. Analyst 106, 992-998 (1981)
Pareles, S.R., Collins, G.J., Rosen, J.D.: Analysis of T-2 toxin (and HT-2 toxin) by mass fragmentography. J. Agric. Food Chem. 24, 872-875 (1976)
Pathre, S.V., Mirocha, C.J.: Assay methods for trichothecenes and review of their natural occurrence. In: Mycotoxins in Human and Animal Health (eds. J.V. Rodricks, C.W. Hesseltine, M.A. Mehlman), pp. 229-253. Park Forest South, IL: Pathotox Publishers 1977
Pathre, S.V., Mirocha, C.J.: Trichothecenes: Natural occurrence and potential hazard. J. Am. Oil Chem. Soc. 56, 820-823 (1979)
Pathre, S.V., Mirocha, C.J., Christensen, C.M., Behrens, J.: Monoacetoxyscirpenol. A new mycotoxin produced by *Fusarium roseum Gibbosum*. J. Agric. Food Chem. 24, 97-103 (1976)
Patterson, D.S.P., Matthews, J.G., Shreeve, B.J., Roberts, B.A., McDonald, S.M., Hayes, A.W.: The failure of trichothecene mycotoxins and whole cultures of *Fusarium tricinctum* to cause experimental haemorrhagic syndromes in calves and pigs. Vet. Record 105, 252-255 (1979)
Pearson, A.J., Ong, C.W.: Trichothecene analogues. Total synthesis of 12, 13-epoxy-14-methoxytrichothecene via organoiron complexes. J. Am. Chem. Soc. 103, 6686-6690 (1981)
Pearson, A.W.: Biochemical changes produced by Fusarium T-2 toxin in the chicken. Res. Vet. Sci. 24, 92-97 (1978)
Penn, J.D.: Mycotoxin testing: A chemical controversy. Clin. Chem. News 8, 1-3 (1982)
Pestka, J.J., Lee, S.C., Lau, H.P., Chu, F.S.: Enzyme-linked immunosorbent assay for T-2 toxin. J. Am. Oil Chem. Soc. 58, 940A-944A (1981)
Petrie, L., Robb, J., Stewart, A.F.: The identification of T-2 toxin and its association with a haemorrhagic syndrome in cattle. Vet. Record 101, 326 (1977)
Pienaar, J.G., Kellerman, T.S., Marasas, W.F.O.: Field outbreaks of leukoencephalomalacia in horses consuming maize infected by *Fusarium-verticillioides (= F. moniliforme)* in South Africa. J. South Afr. Vet. Assoc. 52, 21-24 (1981)
Pokrovsky, A.A., Tutelyan, V.A., Kravenchenko, L.V.: On the mechanism of toxic effect of mycotoxin from *Fusarium sporotrichiella*. Proc. Acad. Med. Sci. (SSSR) 22, 581-595 (1976)
Prior, M.G.: Evaluation of brine shrimp (*Artemia salina*) larvae as a bioassay for mycotoxins in animal feedstuffs. Can. J. Comp. Med. 43, 352-355 (1979)
Puls, R., Greenway, J.A.: Fusariotoxicosis from barley in British Columbia. II. Analysis and toxicity of suspected barley. Can. J. Comp. Med. 40, 16-19 (1976)
Rajendran, M.P., Hissain, M.J., Ramani, K.: A note on stachybotryotoxicosis in Tamil Nadu. Indian Vet. J. 52, 233-235 (1975)

Reiss, J.: Mycotoxin poisoning of *Allium cepa* root tips. II. Reduction of mitotic index and formation of chromosomal aberrations and cytological abnormalities by patulin, rubratoxin B and diacetoxyscirpenol. Cytologia 39, 703-708 (1974)

Reiss, J.: Insecticidal and larvicidal activities of the mycotoxins aflatoxin B_1, rubratoxin B, patulin and diacetoxyscirpenol towards *Drosophila melanogaster*. Chem. Biol. Interact. 10, 339-343 (1975)

Richard, J.L., Cysewski, S.J., Pier, A.C., Booth, G.D.: Comparison of effects of dietary T-2 toxin on growth, immunogenic organs, antibody formation and pathologic changes in turkeys and chickens. Am. J. Vet. Res. 39, 1674-1679 (1978)

Robison, T.S., Mirocha, C.J., Kurtz, H.J., Behrens, J.C., Weaver, G.A., Chi, M.S.: Distribution of tritrium-labeled T-2 toxin in swine. J. Agric. Food Chem. 27, 1411-1413 (1979a)

Robison, T.S., Mirocha, C.J., Kurtz, H.J., Behrens, J.C., Chi, M.S., Weaver, G.A., Nystrom, S.D.: Transmission of T-2 toxin into bovine and porcine milk. J. Dairy Sci. 62, 637-641 (1979b)

Rosenstein, Y., Lafarge-Frayssinet, C., Lespinats, G., Loisillier, F., Lafont, P., Frayssinet, C.: Immunosuppressive activity of *Fusarium* toxins - Effects on antibody synthesis and skin grafts of crude extracts, T-2 toxin and diacetoxyscirpenol. Immunology 36, 111-117 (1979)

Rosenstein, Y., Kretschmer, R.R., Lafarge-Frayssinet, C.: Effect of *Fusarium* toxins, T-2 toxin and diacetoxyscirpenol on murine T-independent immune responses. Immunology 44, 555-560 (1981)

Rukmini, C., Bhat, R.V.: Occurrence of T-2 toxin in *Fusarium* infested sorghum from India. J. Agric. Food Chem. 26, 647-649 (1978)

Rukmini, C., Prasad, J.S., Rao, K.: Effects of feeding T-2 toxin to rats and monkeys. Food Cosmet. Toxicol. 18, 267-269 (1980)

Rüsch, M.E., Stähelin, H.: Über einige biologische Wirkungen des Cytostaticum Verrucarin A. Arzneim. Forsch 15, 893-897 (1965)

Saito, M., Tatsuno, T.: Toxins of *Fusarium nivale*. In: Microbial Toxins (eds. S. Kadis, A. Ciegler, S.J. Ajl), Vol VII, pp. 293-316. New York: Academic Press 1971

Saito, M., Enomoto, M., Tatsuno, T.: Radiomimetic biological properties of the new scirpene metabolites of *Fusarium nivale*. Gann 60, 599-603 (1969)

Saito, M., Horiuchi, T., Ohtsubo, K., Hatanaka, Y., Ueno, Y.: Low tumor incidence in rats with long-term feeding of fusarenon-X, a cytotoxic trichothecene produced by *Fusarium nivale*. Jap. J. Exp. Med. 50, 293-302 (1980)

Sano, A., Asabe, Y., Takitani, S., Ueno, Y.: Fluorodensitometric determination of trichothecene mycotoxins with nicotinamide and 2-acetylpyridine on a silica gel layer. J. Chromatogr. 235, 257-265 (1982)

Sato, N., Ueno, Y., Enomoto, M.: Toxicological approaches to the toxic metabolites of *Fusaria*. VIII. Acute and subacute toxicities of T-2 toxin in cats. Jap. J. Pharmacol. 25, 263-270 (1975)

Schiller, C.M., Yagen, B.: Inhibition of mitochondrial respiration by trichothecene toxins from *Fusarium sporotrichioides*. Fed. Proc. 40, 1579 (1981)

Schindler, D.: Two classes of inhibitors of peptidyl transferase activity in eukaryotes. Nature 249, 38-41 (1974)

Schindler, D., Grant, P., Davies, J.: Trichdermin resistance-mutation affecting eukaryotic ribosomes. Nature 248, 535-536 (1974)

Schmidt, R., Ziegenhagen, E., Dose, K.: High-performance liquid chromatography of trichothecenes. I. Detection of T-2 toxin and HT-2 toxin. J. Chromatogr. 212, 370-373 (1981)

Schmidt, R., Przybylski, M., Dose, K.: Identification of acetyl-T-2 toxin, a trichothecene, in moldy rice by HPLC and FIMS. Fresenius Z. Anal. Chem. 311, 402-403 (1982)

Schneider, D.J., Marasas, W.F.O., Kuys, J.C.D., Kriek, N.P.J., van Schalkwyk, G.C.: A field outbreak of suspected stachybotryotoxicosis in sheep. J. South Afr. Vet. Assoc. 50, 73-81 (1979)

Schoental, R.: The possible public health significance of *Fusarium* toxins. Proc. 3rd Mtg. Mycotoxins in Animal Disease, pp. 67-70. Weybridge: Ministery of Agriculture, Fisheries, and Food; Agricultural Advisory Service 1979a

Schoental, R.: Moldy grain and the aetiology of pellagra: The role of toxic metabolites of *Fusarium*. Biochem. Soc. Trans. 8, 147-150 (1980a)

Schoental, R.: Relationships of *Fusarium* mycotoxins to disorders and tumors associated with alcoholic drinks. Nutr. Cancer 2, 88-92 (1980b)

Schoental, R., Joffe, A.Z.: Lesions induced in rodents by extracts from cultures of *Fusarium poae* and *sporotrichioides*. J. Pathol. 112, 37-42 (1974)

Schoental, R., Joffe, A.Z., Yagen, B.: Irreversible depigmentation of dark mouse hair by T-2 toxin (a metabolite of *Fusarium sporotrichioides*) and by calcium pantothenate. Experientia 34, 763-764 (1978)

Schoental, R., Joffe, A.Z., Yagen, B.: Caridovascular lesions and various tumors found in rats given T-2 toxin, a trichothecene metabolite of *Fusarium*. Cancer Res. 39, 2179-2189 (1979)

Scott, P.M., Harwig, J., Blanchfield, B.J.: Screening *Fusarium* strains isolated from overwintered Canadian grains for trichothecenes. Mycopathologia 72, 175-180 (1980)

Scott, P.M., Lau, P.-Y., Kanhere, S.R.: Gas chromatography with electron capture and mass spectrometric detection of deoxynivalenol in wheat and other grains. J. Assoc. Off. Anal. Chem. 64, 1364-1371 (1981)

Seagrave, S.: Yellow rain. New York: Evans 1981

Siegfried, R.: *Fusarium* - toxine. Naturwissenschaften 64, 274 (1977)

Sigg, H.P., Mauli, R., Flury, E., Hauser, D.: Die Konstitution von Diacetoxyscirpenol. Helv. Chim. Acta 48, 962-988 (1965)

Siriwardana, T.M.G., Lafont, P.: New sensitive biological assay for 12,13-epoxytrichothecenes. Appl. Environ. Microbiol. 35, 206-207 (1978)

Smalley, E.B., Strong, F.M.: Toxic trichothecenes. In: *Mycotoxins* (ed. I.F.H. Purchase), pp. 199-228. Amsterdam: Elsevier 1974

Snyder, W.C., Hansen, H.N.: The species concept in *Fusarium* with reference to discolor and other sections. Am. J. Bot. 32, 657-666 (1945)

Sorsa, M., Linnainmaa, Penttila, M., Ilus, T.: Evaluation of the mutagenicity of epoxytrichothecene mycotoxins in *Drosophila melanogaster*. Hereditas 92, 163-165 (1980)

Stafford, M.E., McLaughlin, C.S.: Trichodermin, a possible inhibitor of the termination process of protein synthesis. J. Cell Physiol. 82, 121-128 (1973)

Stähelin, V.H., Kalberer-Rüsch, M.E., Signer, E., Lazâry, S.: Über einige biologische Wirkungen des Cytostaticum Diacetoxyscirpenol. Arzneim. Forsch. 18, 989-994 (1968)

Stahr, H.M., Ross, P.F., Hyde, R.W., Obioha, W.: Scirpene toxin analysis of feed associated with animal toxication. Appl. Spectroscopy 32, 167-174 (1978)

Stahr, H.M., Kraft, A.A., Schuh, M.: The determination of T-2 toxin, diacetoxyscirpenol and deoxynivalenol in foods and feeds. Appl. Spectroscopy 33, 294-297 (1979)

Stanford, G.K., Hood, R.D., Hayes, A.W.: Effect of prenatal administration of T-2 toxin to mice. Res. Commun. Chem. Path. Pharmacol. 10, 743-746 (1975)

Steele, W.J., Kochanski, J.W., DeMaggio, G.M.: Mechanisms of inhibition of protein synthesis in vivo by the trichothecene mycotoxins diacetoxyscirpenol, trichodermol and trichodermin. Fed. Proc. 40, 901 (1981)

Steyn, P.S., Vleggaar, R., Rabie, C.J., Kriek, N.P.J., Harington, J.S.: Trichothecene mycotoxins from *Fusarium sulphureum*. Phytochemistry 17, 949-951 (1978)

Still, W.C., Ohmizu, H.: Synthesis of verrucarin A. J. Org. Chem. 46, 5242-5244 (1981)
Szathmary, C., Galacz, J., Vida, L., Alexander, G.: Capillary gas chromatographic-mass spectrometric determination of some mycotoxins causing fusariotoxicoses in animals. J. Chromatogr. 191, 327-331 (1980)
Szigeti, G.: Cited in Ueno (1980a). Ph.D. thesis, Miskolc, Hungary 1976
Takitani, S., Asabe, Y., Kato, T., Suzuki, M., Ueno, Y.: Spectrodensitometric determination of trichothecene mycotoxins with 4-(p-nitrobenzyl) pyridine on silica gel layer. J. Chromatogr. 172, 335-342 (1979)
Tamm, Ch.: Chemistry and biosynthesis of trichothecenes. In: Mycotoxins in Human and Animal Health (eds. J.V. Rodricks, C.W. Hesseltine, M.A. Mehlman), pp. 209-228. Park Forest South, IL: Pathotox. Publishers 1977
Tate, W.P., Caskey, C.T.: Peptidyl transferase inhibition by trichodermin. J. Biol. Chem. 248, 7970-7972 (1973)
Tatsuno, T.: Toxicologic research on substances from *Fusarium nivale*. Cancer Res. 28, 2393-2396 (1968)
Tatsuno, T., Saito, M., Enomoto, M., Tsunoda, H.: Nivalenol, a toxic principle of *Fusarium nivale*. Chem. Pharm. Bull. 16, 2519-2520 (1968)
Tatsuno, T., Fujimoto, Y., Morita, Y.: Toxicological research on substances from *Fusarium nivale*. III. The structure of nivalenol and its monoacetate. Tetrahedron Lett. 33, 2823-2826 (1969)
Teodori, L., Barlogie, B., Drewinko, B., Swartzendruber, D., Mauro, F.: Reduction of 1-β-D-arabinofuranosylcytosine and adriamycin cytotoxicity following cell cycle arrest by anguidine. Cancer Res. 41, 1263-1270 (1981)
Tidd, B.K.: Phytotoxic compounds produced by *F. equiseti*. Part III. NMR spectra. J. Chem. Soc. Sect. C Org. Chem. 218-220 (1967)
Traxler, P., Tamm, Ch.: Die Struktur des Antibioticums Roridin H. Helv. Chim. Acta 53, 1846-1869 (1970)
Traxler, P., Zürcher, W., Tamm, Ch.: Die Struktur des Antibioticums Roridin E. Helv. Chim. Acta 53, 2071-2085 (1970)
Trenholm, H.L., Cochrane, W.P., Cohen, J., Elliot, J.I., Farnworth, E.R., Friend, D.W., Hamilton, R.M.G., Neish, G.A., Standish, J.F.: Survey of vomitoxin contamination of the 1980 white winter wheat crop in Ontario, Canada. J. Am. Oil Chem. Soc. 58, 992A-994A (1981)
Tsunoda, H., Toyazaki, N., Morooka, N., Nakano, N., Yoshiyama, H., Okubo, K., Isoda, M.: Researches on the microorganisms which deteriorate the stored cereals and grains (34). Detections of injurious strains and properties of their toxic substance of scab *Fusarium* blight grown on wheat. Report of the Food Res. Inst. Japan, Dept. of Agric., Nat. Inst. of Health, Tokyo 23, 89-116 (1968)
Tullock, P.H.: CMI mycological papers 42, 130. Kew, Surry: Commonwealth Mycolog. Inst. 1972
Tulshian, D.B., Fraser-Reid, B.: A synthetic route to the C4 octadienic esters of trichothecenes from D-glucose. J. Am. Chem. Soc. 103, 474-475 (1981)
Ueno, Y.: Mode of action of trichothecenes. Pure Appl. Chem. 49, 1737-1745 (1977)
Ueno, Y.: Trichothecene mycotoxins: mycology, chemistry, and toxicology. Adv. Nutr. Res. 3, 301-353 (1980a)
Ueno, Y.: Toxicological evaluation of trichothecene mycotoxins. In: Natural Toxins (eds. D. Baker, T. Wadstrom), pp. 663-671. New York: Pergamon 1980b
Ueno, Y., Fukushima, K.: Inhibition of protein and DNA synthesis in Ehrlich ascites tumor by nivalenol, a toxic principle of *Fusarium-nivale*-growing rice. Experientia 24, 1032-1033 (1968)
Ueno, Y., Matsumoto, H.: Inactivation of some thiol-enzymes by trichothecene mycotoxins from *Fusarium* species. Chem. Pharm. Bull. 23, 2439-2442 (1975)

Ueno, Y., Hosoya, M., Morita, Y., Ueno, I., Tatsuno, T.: Inhibition of the protein synthesis in rabbit reticulocytes by nivalenol, a toxic principle isolated from *Fusarium nivale*-growing rice. J. Biochem. (Tokyo) 64, 479-485 (1968)

Ueno, Y., Ueno, I., Tatsuno, T., Ohokubo, K., Tsunoda, H.: Fusarenon-X, a toxic principle of *Fusarium nivale*-culture filtrate. Experientia 25, 1062 (1969)

Ueno, Y., Ishikawa, Y., Saito-Amakai, K., Tsunoda, H.: Environmental factors influencing the production of fusarenon-X, a cytotoxic mycotoxin of *Fusarium nivale* Fr2B. Chem. Pharm. Bull. (Tokyo) 18, 304-312 (1970a)

Ueno, Y., Ishikawa, Y., Amakai, K., Nakajima, M., Saito, M., Enomoto, M., Ohtsubo, K.: Comparative study on skin-necrotizing effect of scirpene metabolites of *Fusaria*. Jap. J. Exp. Med. 40, 33-38 (1970b)

Ueno, Y., Ishikawa, Y., Nakajima, M., Sakai, K., Ishii, K., Tsunoda, H., Saito, M., Enomoto, M., Ohtsubo, K., Umeda, M.: Toxicological appraoches to the metabolites of *Fusaria*. I. Screening of toxic strains. Jap. J. Exp. Med. 41, 257-272 (1971a)

Ueno, Y., Ueno, I., Iitoi, Y., Tsunoda, H., Enomoto, M., Ohtsubo, K.: Toxicological approaches to the metabolites of *Fusaria*. III. Acute toxicity of fusarenon-X. Jap. J. Exp. Med. 41, 521-539 (1971b)

Ueno, Y., Ishii, K., Sakai, K., Kanaeda, S., Tsunoda, H., Tanaka, T., Enomoto, M.: Toxicological approaches to the metabolites of *Fusaria*. IV. Microbial survey on "Bean-hull poisoning of Horses" with the isolation of toxic trichothecenes neosolaniol and T-2 toxin of *Fusarium solani* M-1-1. Jap. J. Exp. Med. 42, 187-203 (1972a)

Ueno, Y., Sato, N., Ishii, K., Sakai, K., Enomoto, M.: Toxicological approaches to the metabolites of *Fusaria*. V. Neosolaniol, T-2 toxin and butenolide, toxic metabolites of *F. sporotrichioides* NRRL 3510 and *F. poae* 3287. Jap. J. Exp. Med. 42, 461-472 (1972b)

Ueno, Y., Sato, N., Ishii, K., Sakai, K., Tsunoda, H., Enomoto, M.: Biological and chemical detection of trichothecene mycotoxins of *Fusarium* species. Appl. Microbiol. 25, 699-704 (1973a)

Ueno, Y., Nakajima, M., Sakai, K., Ishii, K., Sato, N., Shimada, N.: Comparative toxicology of trichothec mycotoxins: Inhibition of protein synthesis in animal cells. J. Biochem. (Tokyo) 74, 285-296 (1973b)

Ueno, Y., Ishii, K., Sato, N., Ohtsubo, K.: Toxicological approaches to the metabolites of *Fusaria*. VI. Vomiting factor from moldy corn infected with *Fusarium* spp. Jap. J. Exp. Med. 44, 123-127 (1974)

Ueno, Y., Sawano, M., Ishii, K.: Production of trichthecene mycotoxins by *Fusarium* species in shake culture. Appl. Microbiol. 30, 4-9 (1975)

Ueno, Y., Kubota, K., Ito, T., Nakumura, Y.: Mutagenicity of carcinogenic mycotoxins in *Salmonella typhimurium*. Cancer Res. 38, 3536-3542 (1978)

Vesonder, R.F., Ciegler, A.: Natural occurrence of vomitoxin in Austrian and Canadian corn. Eur. J. Appl. Microbiol. Biotechnol. 8, 237-240 (1979)

Vesonder, R.F., Ciegler, A., Jensen, A.H.: Isolation of the emetic principle from *Fusarium*-infected corn. Appl. Microbiol. 26, 1008-1010 (1973)

Vesonder, R.F., Ciegler, A., Jensen, A.H., Rohwedder, W.K., Weisleder, D.: Co-identity of the refusal and emetic principle from *Fusarium*-infected corn. Appl. Environ. Microbiol. 31, 280-285 (1976)

Vesonder, R.F., Ciegler, A., Rogers, R.F., Burbridge, K.A., Bothast, R.J., Jensen, A.H.: Survey of 1977 crop year preharvest corn for vomitoxin. Appl. Environ. Microbiol. 36, 885-888 (1978)

Vesonder, R.F., Ciegler, A., Rohwedder, W.K., Eppley, R.: Re-examination of 1972 midwest corn for vomitoxin. Toxicon 17, 658-660 (1979a)

Vesonder, R.F., Ciegler, A., Burmeister, H.R., Jensen, A.H.: Acceptance by swine and rats of corn amended with trichothecenes. Appl. Environ. Microbiol. 38, 344-346 (1979b)

Vesonder, R.F., Ellis, J.J., Rohwedder, W.K.: Swine refusal factors by *Fusarium* strains and identified as trichothecenes. Appl. Environ. Microbiol. 41, 323-324 (1981a)

Vesonder, R.F., Ellis, J.J., Rohwedder, W.K.: Elaboration of vomitoxin and zearalenone by *Fusarium* isolates and the biological activity of *Fusarium*-produced toxins. Appl. Environ. Microbiol. 42, 1132-1134 (1981b)

Vesonder, R.F., Ellis, J.J., Burmeister, H.R.: Production of vomitoxin and zearalenone by *Fusarium*; microbial activity of T-2 toxin, diacetoxyscirpenol, and vomitoxin: Toxicokinetics of T-2 toxin in swine and cattle. Phytopathology 71, 910 (1981c)

Vesonder, R.F., Ellis, J.J., Kwolek, F., De Marini, J.: Production of vomitoxin on corn by *Fusarium graminearum* NRRL 5883 and *Fusarium roseum* NRRL 6101. Appl. Environ. Microbiol. 43, 967-970 (1982)

Wallace, E.M., Pathre, S.V., Mirocha, C.J., Robison, T.S., Fenton, S.W.: Synthesis of radiolabeled T-2 toxin. J. Agric. Food Chem. 25, 836-838 (1977)

Weaver, G.A., Kurtz, H.J., Mirocha, C.J., Bates, F.Y., Behrens, J.C., Robison, T.S., Gipp, W.F.: Mycotoxin-induced abortions in swine. Can. Vet. J. 19, 72-74 (1978a)

Weaver, G.A., Kurtz, H.J., Mirocha, C.J., Bates, F.Y., Behrens, J.C.: Acute toxicity of the mycotoxin diacetoxyscirpenol in swine. Can. Vet. J. 19, 267-271 (1978b)

Weaver, G.A., Kurtz, H.J., Mirocha, C.J., Bates, F.Y., Behrens, J.C., Robison, T.S.: Effect of T-2 toxin on porcine reproduction. Can. Vet. J. 19, 310-314 (1978c)

Weaver, G.A., Kurtz, H.J., Bates, F.Y., Chi, M.S., Mirocha, C.J., Behrens, J.C., Robison, T.S.: Acute and chronic toxicity of T-2 mycotoxin in swine. Vet. Rec. 103, 531-535 (1978d)

Weaver, G.A., Kurtz, H.J., Mirocha, C.J., Bates, F.Y., Behrens, J.C., Robison, T.S., Swanson, S.P.: The failure of purified T-2 mycotoxin to produce hemorrhaging in dairy cattle. Can. Vet. J. 21, 210-213 (1980)

Weaver, G.A., Kurtz, H.J., Bates, F.Y., Mirocha, C.J., Behrens, J.C., Hagler, W.M.: Diacetoxyscirpenol toxicity in pigs. Res. Vet. Sci. 31, 131-135 (1981)

Wehner, F.C., Marasas, W.F.O., Thiel, P.G.: Lack of mutagenicity to *Salmonella typhimurium* of some *Fusarium* mycotoxins. Appl. Environ. Microbiol. 35, 659-662 (1978)

Wei, C.-M., McLaughlin, C.S.: Structure-function relationship in the 12,13-epoxytrichothecenes. Biochem. Biophys. Res. Commun. 57, 838-844 (1974)

Wei, C.-M., Campbell, I.M., McLaughlin, C.S., Vaughan, M.H.: Binding of trichodermin to mammalian ribosomes and its inhibition by other 12,13-epoxytrichothecenes. Mol. Cell. Biochem. 3, 215-219 (1974)

Wei, R.D., Smalley, E.B., Strong, F.M.: Improved skin test for detection of T-2 toxin. Appl. Microbiol. 23, 1029-1030 (1972)

White, J.D., Matsui, T., Thomas, J.A.: A novel synthesis of tricyclic nucleus of verrucarol. J. Org. Chem. 46, 3376-3378 (1981)

Wilson, C.A., Everard, D.M., Schoental, R.: Blood pressure changes and cardiovascular lesions found in rats given T-2 toxin, a trichothecene secondary metabolite of certain *Fusarium* microfungi. Toxicol. Lett. 10, 35-40 (1982)

Wilson, L., Bamburg, J.R., Mizel, S.B., Grisham, L.M., Creswell, K.M.: Interaction of drugs with microtubule proteins. Fed. Proc. 33, 158-166 (1974)

Wollenweber, H.W., Reinking, O.A.: Die Fusarien. Berlin: Parey 1935

Woronin, M.: Über das "Taumel-Getreide" in Süd-Ussurien. Bot. Z. 49, 81-93 (1891)

Wyatt, R.D., Harris, J.R., Hamilton, P.B., Burmeister, H.R.: Possible outbreaks of Fusariotoxicosis in avians. Avian Dis. 16, 1123-1130 (1972a)

Wyatt, R.D., Weeks, B.A., Hamilton, P.B., Burmeister, H.R.: Several oral lesions in chickens caused by ingestion of dietary fusariotoxin T-2. Appl. Microbiol. 24, 251-257 (1972b)

Wyatt, R.D., Colwell, W.M., Hamilton, P.B., Burmeister, H.R.: Neural disturbances in chickens caused by dietary T-2 toxin. Appl. Microbiol. 26, 757-761 (1973)

Wyatt, R.D., Doerr, J.A., Hamilton, P.B., Burmeister, H.R.: Egg production, shell thickness, and other physiological parameters of laying hens affected by T-2 toxin. Appl. Microbiol. 29, 641-645 (1975a)

Wyatt, R.D., Hamilton, P.B., Burmeister, H.R.: Altered feathering of chicks caused by T-2 toxin. Poultry Sci. 54, 1042-1045 (1975b)

Yagen, B., Joffe, A.Z.: Screening of toxic isolates of *Fusarium poae* and *Fusarium sporotrichioides* involved in causing alimentary toxic aleukia. Appl. Environ. Microbiol. 32, 423-427 (1976)

Yagen, B., Horn, P., Joffe, A.Z., Cox, R.H.: Isolation and structural elucidation of a novel sterol metabolite of *Fusarium sporotrichioides* 921. J. Chem. Soc. Perkin I 2914-2917 (1980)

Yoshizawa, T., Morooka, N.: Deoxynivalenol and its monoacetate. New mycotoxins from *Fusarium roseum* and moldy barley. Agric. Biol. Chem. 37, 2933-2934 (1973)

Yoshizawa, T., Marooka, N.: Biological modifications of trichothecene mycotoxins: Acetylation and deacetylation of deoxynivalenols by *Fusarium* spp. Appl. Microbiol. 29, 54-58 (1975)

Yoshizawa, T., Morooka, N.: Trichothecenes from mold infected cereals in Japan. In: Mycotoxins in Human and Animal Health (eds. J.V. Rodricks, C.W. Hesseltine, M.A. Mehlman), pp. 309-321. Park Forest South, IL: Pathotox Publishers 1977

Yoshizawa, T., Onomoto, C., Morooka, N.: Microbial acetyl conjugation of T-2 toxin and its derivatives. Appl. Environ. Microbiol. 39, 962-966 (1980a)

Yoshizawa, T., Swanson, S.P., Mirocha, C.J.: T-2 metabolites in the excreta of broiler chickens administered ^{3}H-labeled T-2 toxin. Appl. Environ. Microbiol. 39, 1172-1177 (1980b)

Yoshizawa, T., Swanson, S.P., Mirocha, C.J.: In vitro metabolism of T-2 toxin in rats. Appl. Environ. Microbiol. 40, 901-906 (1980c)

Yoshizawa, T., Mirocha, C.J., Behrens, J.C., Swanson, S.P.: Metabolic fate of T-2 toxin in a lactating cow. Food Cosmet. Toxicol. 19, 31-39 (1981)

Young, L.G., Vesonder, R.F., Funnell, H.S., Simons, I.: Moldy corn in diets of swine. J. Anim. Sci. 52, 1312-1318 (1981)

Zürcher, W., Tamm, Ch.: Verrucarins and roridins. XIII. Isolation of 2'-dehydroverrucarin A as a metabolite of *Myrothecium roridum*. Helv. Chim. Acta 49, 2594-2597 (1966)

The Scrapie Agent: A Unique Self-Replicating Pathogen

T. L. German and R. F. Marsh

A. Introduction

Slow virus infections can be divided into those caused by conventional and those caused by unconventional agents. The latter group comprises the subacute spongiform encephalopathies which includes scrapie, kuru, Creutzfeldt-Jakob disease, transmissible mink encephalopathy and chronic wasting disease of captive mule deer. These pathogens are called unconventional for the following reasons: (1) they elicit only degenerative tissue lessions, there is no inflammatory reaction; (2) no specific antibody has been detected in infected hosts nor in animals "immunized" with infectious tissue extracts; (3) viral particles are not seen in affected tissue; (4) they do not replicate in vitro; and (5) infectivity is highly resistant to inactivation with ionizing or ultraviolet irradiation.

It has been apparent for a number of years that our understanding of how these agents replicate and produce disease can only progress by first learning their biochemical nature. Despite claims to the contrary (Anon. 1982), studies on disease processes and agent-host interactions are limited in their ability to expand our knowledge on mechanisms of cell injury or agent variability. Only be recognizing pathogen-specific components can we identify cellular and subcellular areas of replication. How far would we have advanced in our understanding of poliovirus infection if we depended only on in vivo experimentation? In this article we hope to inform the reader by providing a brief background on the biology of the spongiform encephalopathies and a summary of current knowledge on the biochemical nature of the scrapie agent. However, our main purpose is to present new theoretical insights on how these unusual pathogens might replicate and offer several experimental methods to test these hypotheses.

B. Biology

I. Natural Diseases

Scrapie is a natural infection of sheep and goats having a minimum incubation period of one year. The disease has been reported in the United States, Great Britain, Germany, France, Hungary, Norway, Poland, Australia, India, Iceland, New Zealand, and Canada. Australia and New Zealand eradicated scrapie by slaughter-

Progress in Molecular and Subcellular Biology, Vol. 8, edited by F.E. Hahn

ing infected and exposed animals. A similar program in the United States has not yet succeeded in eradicating the disease although to date $5 million in State and Federal funds have been expended on indemnity payments.

Transmissible mink encephalopathy is a disease of commercially-reared mink resulting from feeding scrapie-infected tissues. Twelve incidences of the disease have been reported, the last in East Germany in 1975. Chronic wasting disease of captive mule deer has only recently been recognized as belonging to the spongiform encephalopathies (Williams and Young 1980) and little is known of the natural occurrence of disease.

Kuru, a disease of New Guinea natives, was the first human disease recognized as being similar to scrapie (Hadlow 1959), an observation later substantiated by transmission to subhuman primates (Gajdusek et al. 1966). Creutzfeld-Jakob disease is a presenile dementia occurring worldwide with an incidence of about 1 per million population.

II. Pathogenesis

Studies on scrapie infection in experimental mice inoculated peripherally have shown that the agent initially accumulates in reticuloendothelial cells before spreading into the central nervous system (CNS) where it reaches its highest concentration (Eklund et al. 1967). There is no viremia. Recent evidence (Marsh and Hanson 1975; Kimberlin and Walker 1979; Buyukmihci et al. 1980; Fraser 1982) suggests that the agent may be capable of spreading via nerve fibers, similar to rabies virus. Once in the CNS, the agent replicates relatively rapidly reaching titers of 10^{10} LD_{50} per gram of brain tissue by the eigth week post-intracerebral injection in the hamster. Clinical disease is produced by degeneration of nerve cells which is concomitant with agent replication. The disease is invariably fatal.

III. Transmissibility

Scrapie is transmissible to sheep by either direct contact or by exposure to contaminated pastures. The main route of exposure for transmissible mink encephalopathy (Marsh and Hanson 1979) and kuru (Gajdusek 1977) is believed to be via skin abrasions or breaks in mucous membranes. Most cases of Creutzfeld-Jakob disease are sporadic with no known history of exposure, although cases of human to human transmission have been reported (Duffy et al. 1974; Will and Matthews 1982). A familial form of the disease is recognized in which Creutzfeldt-Jakob disease has occurred in one or more family members over several generations. This increased susceptibility is likely due to genetic predisposition and may be analogous to the well documented breed predisposition of sheep to contract scrapie.

There is no evidence that humans are infected by coming into contact with scrapie-infected sheep or sheep tissues (Brown et al. 1979). It is probable that these agents become adapted to individual species over several passages requiring very long periods of time and that they do not readily cross species barriers, especially by natural means of exposure. It is, therefore, more likely that a human would be most easily infected with a human-passaged agent.

C. Biochemical Characterization

I. Purification

Studies on subcellular fractionation of the scrapie agent in brain homogenates have shown that infectivity is tightly bound to the cell membrane (Millson et al. 1971; Semancik et al. 1976). Attempts to solubilize infectivity using tissue disruption and high speed centrifugation (Malone et al. 1978) or detergent extraction (Hunter and Millson 1967) have been only partially successful, since the first method is relatively inefficient and the second often results in some loss of scrapie activity, especially in the presence of ionic detergents (Somerville et al. 1980). Therefore, it appears that membrane integrity is important for either agent stability or for maximum expression of infectivity, the dilemma of scrapie research. Once membranes are reduced to their constituent parts, infectivity is decreased or lost thereby nullifying the only method of agent detection. The first objective in virus characterization is to purify and concentrate. While it has been possible to purify scrapie infectivity with respect to protein, attempts to concentrate infectivity to titers higher than contained in one gram of brain tissue have been uniformly unsuccessful.

II. Evidence for an Essential Protein Component

Every infectious agent, excepting viroids, have protein components. Though scrapie was once hypothesized to be a viroid (Diener 1972), failure of infectivity to survive phenol extraction (Marsh et al. 1974; Ward et al. 1974) indicated that the agent was not identical with plant viroids. In addition to phenol sensitivity, other evidence indicating a protein component includes inactivation of infectivity by proteolytic digestion (Cho 1980; Prusiner et al. 1981a), urea (Hunter et al. 1969), choatropic salts (Millson et al. 1976; Prusiner et al. 1981b), and diethyl pyrocarbonate (McKinley et al. 1981).

III. A Nucleic Acid Component?

The first indication that the scrapie agent was unusual in its nucleic acid content was the finding of very high resistence to ionizing or ultraviolet irradiation (Alper et al. 1966). This was interpreted at the time as suggesting a lack of nucleic acid in the agent (Alper et al. 1967), but later studies on ultraviolet resistence of viroids in crude suspensions (Diener et al. 1974) stimulated a reappraisal toward a nucleic acid of low molecular weight. This possibility was supported by a report on DNase inactivation of a fraction of scrapie infectivity isolated from the front of an SDS-polyacrylamide gel (Marsh et al. 1978). However, this observation has not been repeated (R.F. Marsh and J.S. Semancik unpubl.) and attempts to identify a unique DNA in Creutzfeldt-Jakob brain by nick translation have been unsuccessful (Manuelidis and Manuelidis 1981). Furthermore, scrapie infectivity has been shown to resist treatments known to inactivate nucleic acids; heating at 65°C in the presence of zinc ions, and exposure to psoralens or 0.5M hydroxylamine (Prusiner 1982). While these studies individually provide less than compelling evidence for the absence of a scrapie nucleic acid, in toto they suggest the possibility that the scrapie agent may not require exogenous nucleic acid for replication.

D. Theoretical Considerations for a Self-Regulated Protein Pathogen

In the preceding sections we have reviewed evidence suggesting the possibility that the unconventional slow virus agents do not contain a genomic nucleic acid in the classical sense. They are certainly not conventional nucleoprotein particles which can be purified by standard biochemical techniques. In this section we will attempt to explain how such an agent might replicate while accounting for the existing biochemical data and the observed biological properties of these agents. Speculation of this type is by no means new. There are several early reports in support of the notion that these pathogens are free of nucleic acid (Alper et al. 1966; Pattison and Jones 1967; Alper et al. 1967) and in the light of new evidence the idea has been recently revived (Prusiner 1982). Although some authors have given brief theoretical consideration to how such an agent could persist (Pattison and Jones 1967; Griffith 1967), none have proposed specific mechanisms or described experimental approaches for their verification.

I. General Scheme

The mechanisms that we will propose are not at variance with the general principles of molecular biology in that they are based upon the malfunctioning of regulatory processes known to exist under normal circumstances. The basic requirement of our models, that there exists a gene (or genes) which code for a protein involved at some pointiin regulating its own synthesis, is depicted in Fig. 1. As shown, this control could be exercised by the intervention of such a protein during transcription, RNA processing or during translation. The basic mechanism would be similar in each case. The protein involved binds to a regulatory site on DNA or RNA bringing about a conformational change resulting in either the untimely synthesis of a usually repressed

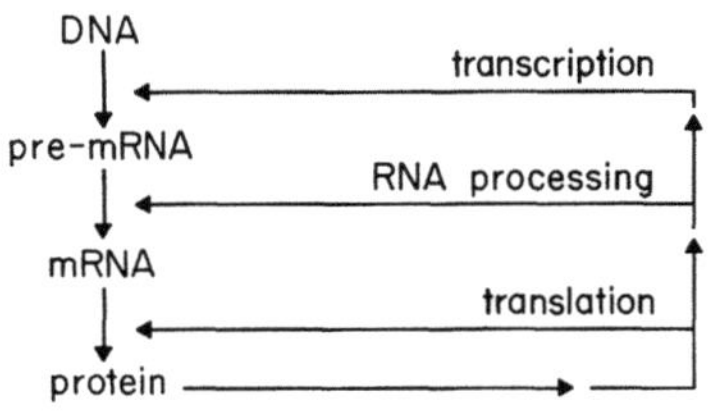

Fig. 1. Generalized scheme of models which we propose for the replication of a self-generating protein agent showing the basic requirement of all our models for a protein encoded in DNA which regulates its own synthesis. This regulation is shown to take place due to intervention of this protein during transcription, RNA processing or during translation

host protein or a slightly altered analogue. The end result in either case is replication of the agent and the production of disease.

II. Transcriptional Control

Transcription of bacterial genes is regulated by the interaction of specific proteins with regulatory sequences located close to transcriptive start sites. This can result in the repression or activation of gene expression. Aminoacyl tRNA synthetases represent an example of a protein which regulates its own synthesis in precisely this manner (Putney and Schimmel 1981). It is likely that the basic physical mechanisms operative in the regulation of bacterial gene expression are utilized in eukaryotic genes. In fact, it has recently been established that there are transciptional control signals involved in a model system employing a eukaryotic protein coding gene (McKnight and Kingsbury 1982).

Our model of transcriptional control (Fig. 2) shows the normal situation in which protein binds to control regions on DNA, alters the structure of downstream regions allowing for transcription of DNA into RNA coding for the regulatory protein (analogous to the aminoacyl synthetase system cited above). The disease state could result from at least two types of improper execution of such a system. If the protein in question were introduced into cells utilizing such a mechanism at the wrong point in time (infection) physiological failure could re-

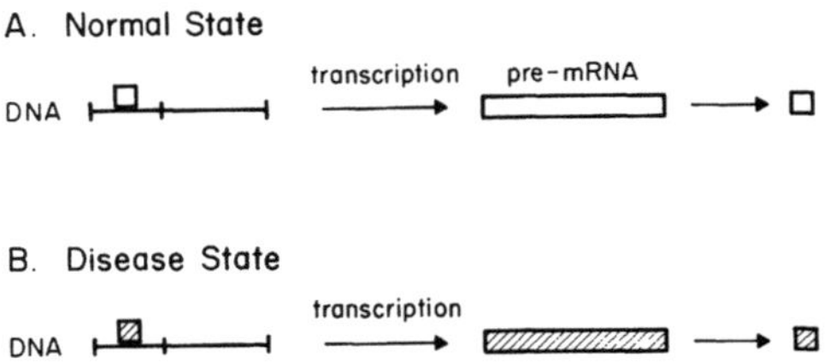

Fig. 2. Diagramatic representation of transcriptional control models. One mechanism (2A) shows a usually repressed normal host protein (*open square*) attaching to regulatory sequences on host DNA (*solid line*) turning on the synthesis of a normal primary transcript (*open rectangle*). Disease would occur if this protein were introduced into a cell in which this transcription was normally turned off. A second mechanism (2B) involves the attachment of an altered protein (*shaded square*) to control regions on host DNA in such a way that transcription produces an altered primary transcript (*shaded rectangle*) coding for the altered protein. Disease would result if this altered protein could not function in place of its normal analogue and since it directs its own synthesis from host nucleic acid the requirements of a self-generating infectious protein are satisfied

sult (disease) and the offending molecule would be synthesized (replication). All somatic cells have the genetic potential to express the same protein constituents, but a selective expression or repression of this information results in a diversity of form and function. If the delicate timing or operation of this expression/repression regulation is disturbed, one possible result would be degenerative disease, especially in as specialized a cell as a neuron.

A second possible mechanism hypothesizes the presence of an altered protein which binds the DNA control region in a slightly different manner from the normal protein. If this resulted in transcription of DNA into RNA coding for this "new" protein, the requirements for a self-generating protein would be met since the altered protein would replace its normal analogue and reproduce itself without the need of transporting its genome between hosts or cells.

III. RNA Processing

In our general scheme, a possible control point is shown to be at the level of RNA processing (Fig. 1). Eukaryotic viruses utilize differential RNA processing to generate multiple protein products from a single transcription unit (Ziff 1980; Darnell 1979). It has been suggested that differential processing of precursor RNA accounts for the production of different mRNAs from identical mouse-amylose genes (Hagenbüchle et al. 1981). Recently, it was shown that multiple mRNAs are generated from calcitonin genes as a consequence of RNA processing events which result in different protein products and seem to occur in a tissue specific fashion (Amara et al. 1982). This phenomenon was first noted during the spontaneous switching of serially transplanted rat medullary thyroid carcinoma lines from states of high to low calcitonin production (Rosenfeld et al. 1981).

Figure 3 shows a diagramatic representation of this process. Since it has been postulated that the protein components of messenger ribonucleoprotein particles are involved in the control of these events (Spirin 1979), it is included in our representation of the normal state (Fig. 3A). In this instance the large primary transcript is cleaved at specific points by nucleases under the control of protein, these cleavage products are then rejoined to form a mRNA for this controlling protein. Here again, the disease state could occur by the untimely implementation of this control mechanism or by the occurrence of a protein that causes misreading of the enzymes involved (Fig. 3B). In the second case, a changed regulatory protein would have arisen as a result of an event modifying either the DNA encoding it, the pre-mRNA or, possibly, the nuclease which does the cutting. The end result of this change is the differential processing of the primary transcripts. That is, instead of fragments a, b, and c of Fig. 3A, fragments d, e and f of Fig. 3B are now obtained. The consequence of this event is the production of messenger d-f (Fig. 3B) rather than a-c (Fig. 3A) and the production of a modified, disease-producing protein in-

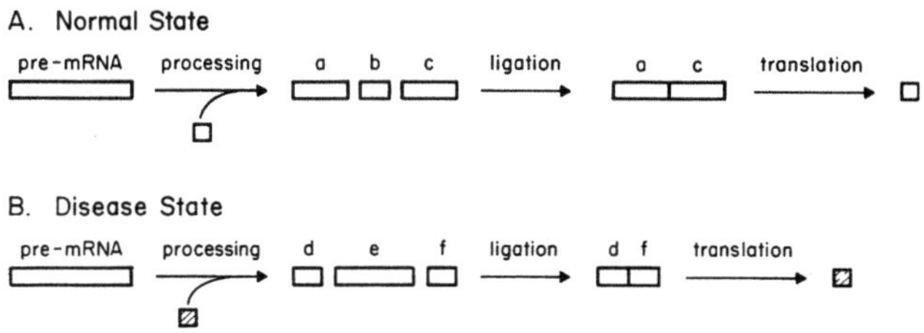

Fig. 3. RNA processing regulatory mechanisms. In the normal state (3A) a primary RNA transcript (*open rectangle*) is cut into several segments (a,b,c) by a nuclease under the control of normal host protein (*open square*). Some of the resulting RNA segments are rejoined to form a functional mRNA (*a–c*) coding for normal protein. Introduction of this normal protein into a cell in which it is usually absent could cause processing of pre-mRNA to take place at an abnormal time. A disease state could also occur (3B) if the regulatory protein was altered (*shaded square*) in such a way that it caused aberrant processing of the normal cell pre-mRNA (*open rectangle*) into fragments (*d, e, f*) and their subsequent rejoining to form mRNA (*d–f*) coding for the altered protein. Disease would occur due to the inability of the altered protein to replace its normal analogue, and the agent would be replicated from host nucleic acid

stead of the normal protein. As in the case in our transcriptional control model, the results are the generation of a transmissible protein from a nucleic acid molecule available in the normal cell.

IV. Translational Control

Translational control affords another possible source for the production of a self-generating protein agent (Fig. 4). Protein is known to control translation in bacteriophage QB (Weber and Konigsberg 1975) and it has been suggested that the association of proteins with globin mRNA and other ribonucleoproteins is due to their involvement in translational control (Vincent et al. 1981). Such control mechanisms involve the binding of protein to control regions on mRNA (Fig. 4A) adjacent to the coding sequences for the structural protein turning on or off the trans-

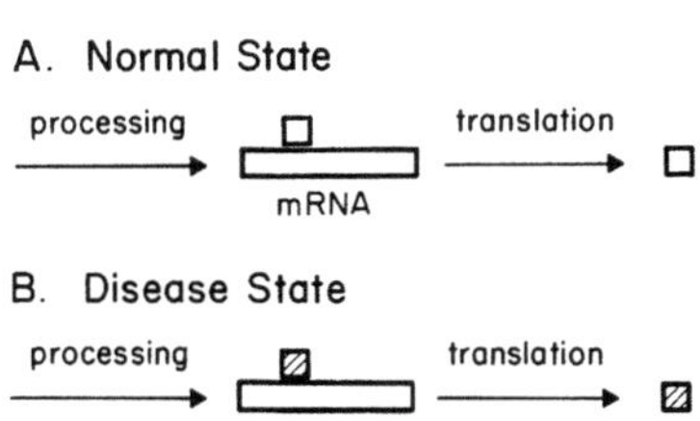

Fig. 4. Translational control mechanisms. In the normal state (4A) a host protein (*open square*) is shown to bind to its mRNA (*open rectangle*) causing a change in tertiary structure allowing translation to begin. If this protein was introduced into cells where the mRNA was present, but not active, protein synthesis could begin at a time that was deliterious to the tissue. Alternatively (2B) altered protein (*shaded square*) could bind the normal mRNA in such a way that the resulting translation product is the altered protein itself. Disease would result from the inability of this altered protein to replace its normal analogue and the protein would be self-generating since it directs its synthesis from a host nucleic acid

lation process by, for example, making ribosome binding sites available by altering the mRNA's tertiary structure. The scenario proposed for the disease state is similar to that of the previously described mechanisms. After processing of the primary transcript into mature mRNA, the intervention of a slightly modified protein could cause a conformational change in the mRNA (Fig. 4B) causing the message to be read differently and be translated into the offending molecule.

V. Relevance

While these models are somewhat speculative they have certain appealing characteristics. They account for the data indicating that these unconventional agents have little or no nucleic acid. In the schemes proposed, a host nucleic acid is used either directly or indirectly to replicate the transmissible agent. In the first case, the protein simply causes its own synthesis to be turned on by utilizing DNA (transcriptional control) or RNA (translational control) already present in the cell. In the indirect case, a genome is created by modification of a pre-existing host nucleic acid by causing misreading during transcription, processing or translation. Our models emphasize protein regulation of these processes, but we do not wish to exclude the possibility that a small subgenomic scrapie nucleic acid may be necessary for recognition and/or binding to specific sites.

These models account for some of the observed biological properties of the agent such as biologic variability of individual isolates, host specificity and long incubation periods. These properties are generally the result of modifications to a pathogen's genome through mutation. In this system, modifications are effected in the pathogen's host range by selection and potentiation of altered proteins having affinity for binding sites on host nucleic acids thereby allowing for selection of biotypes which more easily infect one host than another. Since these proteins must necessarily be the same as host protein (in the case of temporal malfunctioning), or only slightly different, the apparent lack of immunogenicity is explained.

E. Future Directions

Our understanding of the biochemical properties and the mechanism of replication employed by these unconventional virus-like agents is unequivocally dependent upon the purification and identification of the macromolecular structure constituting the transmissible unit. Then, and only then, will it be possible to be certain that the agent does not contain nucleic acid. Since it is likely that the agent contains protein, efforts at this time should be concentrated on the purification and characterization of proteins associated with infectivity. When this is accomplished, replication modes such as those proposed here and elsewhere can be tested. For example, there must exist an mRNA coding for a protein associated with disease. Once the protein involved is identified, its mesenger can be purified by fractionation of RNA from diseased tissue and translated in vitro to find the molecules coating for the protein in question. In vitro systems of transcription and translation can then be used to determine whether these events are subject to control by protein present in the agent and/or normal tissues by examining the transcription, RNA processing events and translation products in the presence or absence of those proteins.

Once the mRNA is identified and charaterized, probes can be made by reverse transcription and used to locate the genes involved in these diseases to answer questions concerning the relateness of these various pathogens. Finally, once the biochemical characteristics of these agents are understood, the search for the involvement of this type of agent in the etiology of other diseases can begin.

References

Alper, T., Haig, D.A., Clarke, M.C.: The exceptionally small size of the scrapie agent. Biochem. Biophys. Res. Commun. 22, 278-284 (1966)

Alper, T., Cramp, W.A., Haig, D.A., Clarke, M.C.: Does the agent of scrapie replicate without a nucleic acid? Nature 214, 764-766 (1967)

Amara, S.G., Jonas, V., Rosenfeld, M., Ong, E., Evans, R.M.: Alternative RNA processing in calcitonin gene expression generates mRNAs encoding different polypeptide products. Nature 298, 240-244 (1982)

Anonymous: Scrapie: Strategies, stalemates, and successes. Lancet 1, 1221-1223 (1982)

Brown, P., Cathala, F., Gajdusek, D.C.: Creutzfeldt-Jakob disease in France: III. Epidemiological study of 170 patients dying during the decade 1968-1977. Ann. Neurol. 6, 438-446 (1979)

Buyukmihci, N., Rorvik, M., Marsh, R.F.: Replication of the scrapie agent in ocular neural tissues. Proc. Natl. Acad. Sci. USA 77, 1169-1171 (1980)

Cho, H.J.: Requirement of a protein component for scrapie infectivity. Intervirol. 14, 213-216 (1980)

Darnel, J.E.: Transcription units for mRNA production in eukaryotic cells and their DNA viruses. Prog. Nucleic Res. Mol. Biol. 22, 327-357 (1979)

Diener, T.O.: Is the scrapie agent a viroid? Nature 235, 218-219 (1972)

Diener, T.O., Schneider, I.R., Smith, D.R.: Potato spindle tuber viroid. XI. A comparison of the ultraviolet light sensitivity of PSTV, tobacco ringspot virus, and its satellite. Virol. 57, 577-581 (1974)

Duffy, P., Wolf, J., Collins, G., DeVoe, A.G., Streeten, B., Cowen, D.: Possible person to person transmission of Creutzfeldt-Jakob disease. N. Engl. J. Med. 290, 692-693 (1974)

Eklund, C.M., Kennedy, R.C., Hadlow, W.J.: Pathogenesis of scrapie virus infection in the mouse. J. Infect. Dis. 117, 15-22 (1967)

Fraser, H.: Neuronal spread of scrapie agent and targeting of lesions within the retino-tectal pathway. Nature 295, 149-150 (1982)

Gajdusek, D.C.: Unconventional viruses and the origin and disappearance of kuru. Science 197, 943-960 (1977)

Gajdusek, D.C., Gibbs, C.J., Jr., Alpers, M.: Experimental transmission of a kurulike syndrome to the chimpanzee. Nature 209, 794-796 (1966)

Griffith, J.S.: Self replication and scrapie. Nature 215, 1043-1044 (1967)
Hadlow, W.J.: Scrapie and kuru. Lancet ii, 289-290 (1959)
Hagenbüchle, O., Tosi, M., Schibler, U., Bovey, R., Wellauer, P.K., Young, R.A.: Mouse liver and salivary gland-amylase mRNAs differ only in 5' non-translated sequences. Nature 289, 643-646 (1981)
Hunter, G.D., Millson, G.C.: Attempts to release the scrape agent from tissue debris. J. Comp. Pathol. 77, 375-383 (1967)
Hunter, G.D., Gibbons, R.A., Kimberlin, R.H., Millson, G.C.: Further studies on the infectivity and stability of extracts and homogenates derived from scrapie affected mouse brains. J. Comp. Pathol. 79, 101-108 (1979)
Kimberlin, R.H., Walker, C.A.: Pathogenesis of mouse scrapie: Dynamics of agent replication in spleen, spinal cord and brain after infection by different routes. J. Comp. Pathol. 89, 551-562 (1979)
Malone, T.G., Marsh, R.F., Hanson, R.P., Semancik, J.S.: Membrane-free scrapie activity. J. Virol. 25, 933-935 (1978)
Manuelidis, L., Manuelidis, E.E.: Search for specific DNAs in Creutzfeldt-Jakob infectious brain fractions using "nick translation". Virology 109, 435-443 (1981)
Marsh, R.F., Hanson, R.P.: Transmissible mink encephalopathy: Infectivity of corneal epithelium. Science 187, 656 (1975)
Marsh, R.F., Hanson, R.P.: On the origin of transmissible mink encephalopathy. In: Slow Transmissible Diseases of the Nervous System (eds. W.J. Hadlow, S. Prusiner), Vol. 1, pp. 451-460. New York: Academic Press 1979
Marsh, R.F., Semancik, J.S., Medappa, K.C., Hanson, R.P., Rueckert, R.R.: Scrapie and transmissible mink encephalopathy: Search for infectious nucleic acid. J. Virol. 13, 993-996 (1974)
Marsh, R.F., Malone, T.G., Semancik, J.S., Lancaster, W.D., Hanson, R.P.: Evidence for an essential DNA component in the scrapie agent. Nature 275, 146-147 (1978)
McKinley, M.P., Masiarz, F.R., Prusiner, S.B.: Reversible chemical modification of the scrapie agent. Science 214, 1259-1261 (1981)
McKnight, S.L., Kingsbury, R.: Transcriptional control signals of a eukaryotic protein-coding gene. Science 217, 316-324 (1982)
Millson, G.C., Hunter, G.D., Kimberlin, R.H.: An experimental examination of the scrapie agent in cell membrane mixtures. II. The association of scrapie activity with membrane fractions. J. Comp. Pathol. 81, 255-265 (1971)
Millson, G.C., Hunter, G.D., Kimberlin, R.H.: The physico-chemical nature of the scrapie agent. In: Slow Virus Diseases of Animals and Man (ed. R.H. Kimberlin), pp. 243-266. Amsterdam New York: Elsevier, North-Holland 1976
Pattison, I.H., Jones, K.M.: The possible nature of the transmissible agent of scrapie. Vet. Rec. 80, 2-9 (1967)
Prusiner, S.B.: Novel proteinaceous infectious particles cause scrapie. Science 216, 136-144 (1982)
Prusiner, S.B., McKinley, M.P., Groth, D.F., Bowman, K.A., Mock, N.I., Cochran, S.P., Masiarz, F.R.: Scrapie agent contains a hydrophobic protein. Proc. Natl. Acad. Sci. USA 78, 6675-6679 (1981a)
Prusiner, S.B., Groth, D.F., McKinley, M.P., Cochran, S.P., Bowman, K.A., Kasper, K.C.: Thiocyanate and hydroxyl ions inactive the scrapie agent. Proc. Natl. Acad. Sci. USA 78, 4606-4610 (1981b)
Putney, S.D., Schimmel, P.: An aminoacyl tRNA synthetase binds to a specific DNA sequence and regulates its gene transcription. Nature 291, 632-635 (1981)
Rosenfeld, M.G., Amara, S.G., Roos, B.A., Ong, E.S., Evans, R.M.: Altered expression of the calcitonin gene associated with RNA polymorphism. Nature 290, 63-65 (1981)

Semancik, J.S., Marsh, R.F., Geelen, J.L.M.C., Hanson, R.P.: Properties of the scrapie agent-endomembrane complex from hamster brain. J. Virol. 18, 693-700 (1976)

Spirin, A.: Messenger ribonucleoproteins (Informosomes) and RNA binding proteins. Mol. Biol. Rep. 5, 53-57 (1979)

Somerville, R.A., Millson, G.C., Kimberlin, R.H.: Sensitivity of scrapie infection to detergents and 2-mercaptoethanol. Intervirology 13, 126-129 (1980)

Vincent, A., Goldenberg, S., Scherrer, K.: Comparisons of proteins associated with duck-globin mRNA and its polyadenylated segment in polyribosomal and repressed free messenger ribonucleoprotein complexes. Eur. J. Biochem. 114, 179-193 (1981)

Ward, R.L., Porter, D.D., Stevens, J.G.: Nature of the scrapie agent: Evidence against a viroid. J. Virol. 14, 1099-1103 (1974)

Weber, K., Konigsberg, W.: Proteins of the RNA phages. In: RNA Phages (ed. N. Zinder), pp. 51-84. Cold Spring Harbor Laboratory 1975

Will, R.G., Matthews, W.B.: Evidence for case to case transmission of Creutzfeldt-Jakob disease. J. Neurol. Neurosurg. Psych. 45, 235-238 (1982)

Williams, E.S., Young, S.: Chronic wasting disease of captive mule deer: A spongiform encephalopathy. J. Wildlife Dis. 16, 89-98 (1980)

Ziff, E.B.: Transcription and RNA processing by the DNA tumor viruses. Nature 287, 491-499 (1980)

Non-Cholinesterase Effects of Anticholinesterases

John J. O'Neill

A. Background

Largely through an interest in developing agricultural pesticides, the I.G. Farbenindustrie in the mid-1930's began a synthetic program under the direction of Gerhard Schrader. Organophosphorus compounds had previously been described with insecticidal properties by Lange but had attracted little attention prior to that time. In the course of synthesis, Schrader became aware of a miosis and other mild symptoms which were ascribable to the action of the organophosphates. His synthetic studies lead to the formation of diisopropylfluorophosphate (DFP) and some 2,000 other compounds of which about 10% were toxic. From animal studies, it was early recognized that their pharmacologic action resembled that of eserine (physostigmine) except that inhibition of cholinesterase was irreversible. Largely through an interest in developing more potent chemical warfare agents on the lines of Schrader, the potent organophosphates Tabun (ethyl N, N-dimethylphosphoramidocyanidate) Sarin (isopropyl-methylphosphonofluoridate) and Soman (pinacolyl-methylphosphonofluoridate) were synthesized in large quantity and their potent toxic action studied. A thorough coverage of this topic may be found in an informative discussion by Professor Bo Holmstedt (1963).

It is generally accepted that acute exposure to high concentrations of Soman, Tabun or other potent organophosphates (OP's) leads to the irreversible inhibition of acetylcholinesterase. It is also assumed that the toxicity which follows a lethal exposure to OP's is due to acetylcholine accumulation and poisoning (Du Bois 1963). The evidence is persuasive that, if left untreated, the immediate cause of death following a lethal exposure in man is due to asphyxia following respiratory paralysis, pulmonary edema and cardiac arrest.

The organophosphate compounds (OP's) are known to inactivate the enzyme acetylcholinesterase from all sources studied including vertebrates other than man, a wide variety of invertebrates and insects. They also inactivate plasma or pseudocholinesterases, the major differences being in rate of inhibition. Most striking is the case of diisopropylphosphorofluoridate in which the plasma enzyme bimolecular inhibitory rate constant is approximately 3 orders of magnitude greater than for acetylcholinesterase. The inhibitory action of OP's is not exclusive for cholinesterases but occurs in a broad category of enzymes, characterized by structural features as *serine proteases*. It is the purpose of the present chapter to offer evidence that organophosphate "anti-

Progress in Molecular and Subcellular Biology, Vol. 8, edited by F.E. Hahn

cholinesterases" react with other enzymes present in brain and nervous tissue, involved in neuronal function. By such interaction, toxic organophosphate compounds may produce undesirable actions in ways other than through acetylcholine poisoning.

It is over thirty years since the report of Schaffer et al. (1956 describing the phosphorylation of the *serine protease*, chymotrypsin by radiolabeled DF^{32}P. This work was based on the earlier finding of Jansen et al. (1949) that DFP reacted with chymotrypsin and trypsin. This breakthrough formed the basis of intensive research at the Army Chemical Center Maryland under the direction of William Summerson, Bernard Jandorf, and Theodore Wagner-Jauregg. Studies were designed to elucidate the chemical nature of the active center in the crystalline α-chymotrypsin and to use that information to study the active center in acetylcholinesterase. Similar research was intensively pursued by Wilson and Nachmansohn at Columbia University, by Cohn and Oosterbaan in the Netherlands, and by the British group at Porton Downs. These combined efforts clearly established that like α-chymotrypsin, Sarin, or GB as it was more commonly referred to, reacted stoichiometrically with acetylcholinesterase and that serine methyl phosphonate was the stable residue. The reaction between α-chymotrypsin and inhibitor yielded an inactive derivative containing a single methyl phosphonyl radical attached to a specifically located serine group in the "active center". The fact that the substitution of a single radical at a specific locus causes total inhibition argues strongly in support of the generally accepted notion that there exists in an enzyme molecule, a single and relatively small peptide sequence endowed with proteolytic or esteratic activity. The partial sequence of the active center of chymotrypsin from the studies of Schaffer et al. (1957) and of Oosterbaan et al. (1958) are shown (Fig. 1)

Fig. 1

The B group in the hydrolysates of phosporylated esterases found

^{32}P (on Ser)
Gly—Asp—Ser—Gly—Gly—Pro—Leu;

the latter

^{32}P (on Ser)
Gly—Asp—Ser—Gly—Glu—Ala.

As far as trypsin is concerned, the most complete sequence reported by Dixon et al. (1958a, 1958b).

NH_2

(Asp Ser Cys Glu Gly Gly Asp Ser Gly Pro Val Cys Ser Gl- Lys),
(each Cys bearing SO_3^-)

Fig. 1. The sequence about the ^{32}P-labeled serine in chymotrypsin. Schaffer et al. (1957) found a gly-gly adjacent to the carboxyl group of serine while Oösterban et al. (1958) found gly-glu. The sequence described by Dixon et al. (1958) is also included

Fig. 2

Labelled enzyme	Hydrolysis method	Amino acid composition or sequence	References
Chymotrypsin-DFP	HCl	P (on SER) GLY—ASP—SER—GLY	Schaffer et al. (1957)
	HCl	P (on SER) ASP—SER—GLY(Glu,Ala,Gly)	Turba and Gundlach (1955)
	"Cotazym"	P (on SER) GLY—ASP—SER—GLY—GLY—PRO—LEU	Oosterbaan et al. (1958a, 1958b)
Chymotrypsin-Sarin	HCl	P (on SER) GLY—ASP—SER—GLY—GLU—ALA(Val)	Schaffer et al. (1956, 1957)
	Papain	P (on SER) GLY—ASP—SER—GLY—GLU—ALA(Val,His,Pro,Leu,Cys,Thr)	Schaffer et al. (1956, 1957)
Trypsin-Sarin	HCl	P (on SER) ASP—SER—GLY	Schaffer et al. (1958)
Trypsin-DFP		NH_2 (on the first ASP); P (on the second SER); SO_3H (on each CYS) ASP—SER—CYS—GLU—GLY—GLY—ASP—SER—GLY—PRO—VAL—CYS—SER—GLY—LYS	Dixon et al. (1958a, 1958b)
	Chymotrypsin	P (on Ser) Gly,Asp,Ser,Gly,Pro,Val,Cys,Ala,Glu,Lys	Dixon et al. (1956b)
	"Cotazym"	P (on Ser) Gly,Asp,Ser,Gly,Pro,Val	Oosterbaan et al. (1956)

Fig. 2 (cont.)

Labelled enzyme	Hydrolysis method	Amino acid composition or sequence	References
Liver ali-esterase-DFP	Pepsin	P GLY—GLU—SER—ALA—GLY—GLY—(GLU,SER)	Jansz et al. (1959b, 1959c)
Butyrocholin-esterase-DFP	Pepsin	P PHE—GLY—GLU—SER—ALA—GLY—(ALA,ALA,SER)	Jansz et al. (1959a)
Thrombin-DFP	HCl	P ASP—SER—GLY and P Asp,Ser,Gly,Glu,Ala	Gladner and Laki (1958)
Phospho-glucomu-tase	HCl or Proteo-lysis	PO_3H_2 ASP—SER—(Gly,Glu) and PO_3H_2 ASP—SER—(Gly,Glu,Ala,Val,Thr,Leu)	Koshland and Erwin (1957);; Koshland et al. (1958) *)
Chymotrypsin-NPA	Pepsin and "Cotazym"	O=C—CH_3 GLY—ASP—SER—GLY—GLY—PRO—LEU	Cohen et al. (1959)

Remarks: P in the peptide structures denotes the phosphorylgroup originating from DFP or Sarin.
GLY-etc. = amino acid residue in established sequence. GLY, = amino acid residue in unknown sequence.
GLY, etc. = sequence and exact number of residues unknown. NPA = *p*-nitrophenyl acetate.

*) Recently C. Milstein and F. Sanger [Biochem. J. 79, 456-469 (1961)] found the sequence THR. ALA. P-SER. HIS-ASP or ($ASP\text{-}NH_2$).

Fig. 2. Amino acid compositions and sequences in the active sites of enzymes

together with that of trypsin. The similarities between the two studies are striking; the minor differences in sequence may be due to the use of different techniques for partial protein hydrolysis i.e., Oosterbaan used proteolytic enzymes, Schaffer employed the concentrated hydrochloric acid method which Sanger employed in the characterization of insulin.

It is highly significant that trypsin, despite its marked differences in substrate specificity from chymotrypsin has an analogous structure about the phosphorylated serine residue. Subsequent analyses revealed an entire group of enzymes now designated as *serine proteases*. The serine proteases may be viewed as a family of enzymes having similar amino acid sequences in the active center but differing in substrate specificity. An early and incomplete list of these proteases, all of which contain serine phosphate in the inactivated state, is shown in Fig. 2. The list now includes the kallikreins, important as bioregulatory serine proteases, kininogenases, elastases, thrombin, plasmin, C-1 protease, bacterial subtilisin, collagenases and others, even the silk worm cocoonase. A recent review by Schachter (1980) in this regard recently appeared. In his article on serine proteases, Kraus (1970) expresses the view that "inhibition of any protease by diisopropylfluorophosphate is generally considered diagnostic for its identification as a *serine* protease".

At the time studies on the structure of active centers in enzymes flourished, Koshland wrote about the role of serine in the catalytic center (1960). He stressed the important contribution made by other amino acids to the specificity of a particular enzyme and in particular the amino acid, histidine. Although at a distance from serine in an amino acid sequence, it may be "in juxta-position at the active site because of three-dimensional coiling of the molecule". This is important because of the evidence of Jandorf et al. (1955) and of Wagner-Jauregg and Hackley (1953) that a histidine residue is essential for catalytic activity and may be involved in the "aging" process following cholinesterase inhibition by Soman and other organophosphates. A recent paper in Nature by Kossiakoff et al. (1980) using neutron diffraction (Fig. 3) identified Histidine^{57} as the catalytic base in trypsin. Interestingly, in the process involving the interaction of Serine^{195} with DFP, one of the isopropyl groups was lost similar to the "aging" phenomenon in vivo.

The importance of these early observations with regard to the reactivity of peptidases and proteases has acquired added significance with the newly discovered neuromodulatory role of endoggenous neuropeptides. At the time when there was an intensive effort to find a therapeutic approach to organophosphate poisoning, the various labs were most interested in defining the mechanism by which acetylcholinesterase was inhibited. This we believed would provide a rational approach for designing effective therapeutic agents. Since acetylcholinesterase(s) were only partially purified, the crystalline serine proteases were selected for study in parallel. As previously indicated, they reacted stoichiometrically (1:1) with organophosphates by covalent bonding to serine, possessed esteratic and peptidase activities

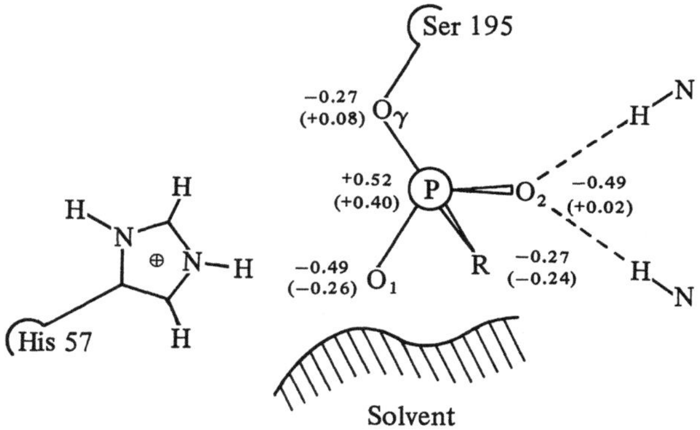

Fig. 3. Schematic diagram showing the orientation of the MIP group with respect to the imidazole of His[57]. The MIP group resembles the deacylation intermediate of the reaction. In the real intermediate, the phosphorus is replaced by a carbon atom and oxygen O-1 by a hydroxyl group. Ligand R corresponds to the amino acid side chain, which in MIP is an isopropyl group. The individual charges on the atoms in the MIP group are differences in charge between corresponding atoms in the MIP group and the tetrahedral intermediate. The charges for the tetrahedral intermediate were obtained from a model reaction study (using methoxide ion and formamide) of the acylation step of the reaction. The charge on the hydroxyl group was assumed to be approximately the same as that on NH_2 in the methoxideformamide intermediate. This assumption is supported by the calculated charge distributions of a hydroxylformamide intermediate. Electrostatic potential energies were calculated using the expression $E = \sum(q_i q_j)/\varepsilon_O\Gamma)$. ε_O, the dielectric constant, was 3.0; Γ is the distance between charges q_i and q_j

which were easily measurable, and, inhibited enzymes could be isolated in crystalline form. The ability of proteases to hydrolyze peptides with specific aminoacid sequences was known but its significance was not considered to be related to organophosphate poisoning at the time. The discovery of peptides in brain and nerve tissue with marked biologic activity mainly through the pioneer work of Hokfelt, Kosterlitz and Hughes, Schally, Guillemin, Goldstein, Leemans, Iversen, Snyder and others, gave new implications regarding serine proteases and the organophosphates.

B. Neuropeptides

The classic studies of Professor Cho Hao Li on the anterior pituitary hormone, ACTH, were originally thought to be largely of interest to endocrinologists. This may also be said of pituitary peptides known to influence vascular or uterine smooth muscle, e.g., vasopressin and oxytocin. Despite earlier evidence by Von Euler and Gaddum that the peptide Substance P had marked pharmacologic effects when applied to smooth muscle, it was the research of Bargmann and collaborators (1967), which initiated the concept of a "peptidergic neuron". According to Phillis (1979) in his discussion of the effects on the CNS of locally applied peptides, prior evidence had established that the hypothalamic and pituitary tissues, when stimulated, were caused to release hormones directly into the blood stream which transported them to the target tissue. In Bargmann's paper, however, evidence was presented to show that while such release did occur, neurons containing peptides were present which formed synapses on epithelial cells of the pars intermedia of the pituitary. This led to the concept of the "peptidergic neuron". It is now appreciated that not only neurosecretory cells with endocrine function but also neurons contain a variety of peptides which they may be caused to release under proper stimulus. Largely through the use of elegant immunocytochemical methods developed by L. Sternberger Hokfelt and collaborators have led the way in describing the distribution of Thyrotropin Releasing Hormone (TRH), Somatostin, Substance P and the enkephalins in nerve cell bodies and axonal processes in brain and spinal cord. With special reference to these peptides Hokfelt indicated that although cell bodies and nerve fibers containing TRH form intrahypothalamic pathways, approximately 75% of TRH is distributed elsewhere in the CNS. As to the functional role of peptides, Hokfelt believes that in the median eminence, TRH and Somatostatin either cause or inhibit the release of hormones into portal vessels affecting the pituitary, and the enkephalins probably have some local effect there. Outside the median eminence, all may serve as transmitters or modulators, mainly of an inhibitory nature. What is perhaps more surprising is that numerous peptides first described in the digestive tract and thought to possess a physiological role locally, are now shown to be widely distributed in neurons throughout regions of brain and spinal cord.

In immunochemistry it is a recognized limitation that antibodies, while possessing group specificity, nevertheless undergo cross-reactivity with other substances possessing a particular amino acid sequence in its composition. With this reservation notwithstanding, it can be shown that CCK-8, the octapeptide derivated from the parent cholecystokinin containing 33 amino acids, is present in neurons of the cerebral cortex, hippocampus, hypothalamus, amygdaloid nucleus and spinal cord (Larsson and Richfield 1979). Cholecystokinin originally found in the duodenum is present in brain at the same concentration per gram of tissue, but since in humans the brain mass (circa 1,500 g) is approximately 30 times that of duodenum, there is considerably more CCK-8 in brain according to Yalow et al. (1981). They showed CCK-8 brain levels were altered significantly dependent on the nutritional state of the animal. It is significant in this regard that Strauss et al. (1978) in Yalow's laboratory have purified an enzyme from brain which cleaves CCK-33 at the arginine-isoleucine bond to form CCK-12 and at the arginine-aspartate linkage to form CCK-8. It is very similar in action to trypsin but distinguishable by its greater group-specificity. There is a synaptosomal localization for CCK-8 in rat brain which is dependent on the presence of calcium ions for release (Emson 1979). CCK-8 has an excitatory action on hippocampal pyramidal cells and a depolarizing action on spinal cord dorsal root terminals and motorneurons not blocked by tetrodotoxin (Phillis 1979).

A second peptide originally found in the intestinal tract, Vasoactive Intestinal Peptide (VIP), is also present in high concentrations in the cerebral cortex and hypothalamus. Like CCK-8, it is releasable from nerve endings by potassium in the presence of calcium ions. Iontophoretic application of VIP to the exposed tissues causes excitation of cortical neurons and depolarization of spinal motor neurons (Phillis 1978).

TRH
LHRH
Somatostatin
Substance P
Neurotensin
Oxytocin
Vasopressin
Neurotensin
Enkephalin
Endorphin
Angiotensin
Bombesin
Cholecystokinin
Gastrin
Vasoactive Intestinal polypeptide
ACTH

Fig. 4. A partial list of peptides known to be present in brain. The list is only a sampling of the many biologically active peptides present or with the potential of being formed in nervous tissue

Peptide-containing neurons were at first thought to be few in number in the CNS and to have a highly specialized endocrine function. It is now known that there is a large number of peptidergic neurons which are not related to hormone releasing factors, and which are widely distributed throughout the central and peripheral nervous systems. Only a partial list of many peptides known to be present in brain and nervous tissue is shown in Fig. 4. The simplest of peptides Thyrotropin Releasing Hormone (TRH):

$$\text{pyro-Glu-His-Pro-NH}_2$$

(pyroglutamyl-histidyl-proline amide)

was identified as the tripeptide by Burgus et al. (1969) and by Bohler et al. (1969).

The synthesis of TRH, a tripeptide appears to be different from the formation of enkephalins, endorphins and dynorphin. They are formed by post-translational cleavage of larger protein molecules. The precise mechanism by which TRH is formed is not known but it may be through an mRNA-directed ribosomal formation or by nonribosomal enzymatic synthesis, analogous to the formation of another ubiquitous tripeptide, glutathione. A major complicating factor to an understanding of its formation is the small amount of TRH formed in tissues and the presence of one or more peptidases which attack the amide as well as the peptide bond, similar to the actions of chymotrypsin. Although a number of inhibitors were described, including those affecting mRNA and tRNA, no mention has been made of the actions or organophosphates on the overall process of synthesis and degradation. However, in an earlier review, Marks (1976) stressed the importance of peptide turnover (synthesis and degradation) of neurosecretory peptides and the presence in brain of a large number of exo- and endopeptidases. Some show unique group specificity in the restricted number of biologically active materials which are cleaved, which accounts for this turnover. The presence of the pyroglutamyl-N-terminal amino acid in TRH is a case in point. The primary site of enzymatic attack in brain appears to be the pyro-Glu-His bond as no free proline was found, presumably due to a pyroglutamyl peptidase (EC 3.4.11.9). It is suggested that in TRH, pyro-Glu-His-Pro-NH2, an aminopeptidase, cleaves the proline and the amide linkage analogous to the actions of chymotrypsin but that the primary cleavage occurs at the pyro-Glu-His bond.

The possibility exists therefore that TRH-turnover could be adversely influenced by organophosphate anticholinesterases. It is known that organophosphates produce significant EEG changes in animals following less than lethal exposure. The intracerebroventricular administration of TRH caused a significant increase in both hippocampal and cortical synchrony (Breese et al. 1975). Iontroporetic application to 20 different brain areas revealed the septal area to be particularly sensitive. The arousal produced antagonized pentobarbital narcosis. Such arousal can be shown to differ from motor activity effects which also induce hippocampal synchrony. In these studies TRH was shown to possess marked tremorogenic effects which may be correlated with cortical EEG-synchrony. What adds significance to these observations is evidence that TRH acts via a cholinergic mechanism (Breese 1975) and of the established cholinergic nature of the septo-hippocampal system.

TRH is able to alter a number of physiological parameters — perhaps the most important in the present context is its effect on respiration. Metcalf and Myers (1976) suggested that TRH-induced tachypnea observed in cats may be responsible for the hypothermia observed in other species. TRH and Neurotensin are more effective blockers of opioid peptides than is the opiate antagonist naloxone. The analeptic effects of TRH are also more effective in blocking opioid peptides than is Naloxone. The analeptic effects of TRH are antagonized by antiocholinergic drugs such as atropine. In addition, anti-muscarinic drugs abolish the actions of TRH in activating the EEG in rabbits and block the increased glucose utilization following TRH-treatment of pentobarbital depressed rats. TRH also antagonizes the decreased cholinergic activity caused by other depressant drugs and Yarborough and Singh (1978) found that TRH enhances the excitatory actions of acetylcholine on individual neurons. Lastly, but not less important is the finding by Malthe Sorenson et al. (1978) that TRH reduces brain ACh *content* but increases ACh *turnover rate* in cortical regions.

Many proteins exist in a pro-enzyme or pro-hormone form. They experience specific post-translational changes in gaining functional activity, mostly enzyme directed (Chretien et al. 1979). Recent studies support the suggestion by Habener that most biologically active peptides as well are also derived from protein precursors of larger molecular weight (Habener et al. 1977). Largely as a result of the elegant studies of Choh Hao Li and collaborators (Chretien and Li 1967), we have learned how the biologically active B-endorphin and Dynorphin as well as leu- and met-enkepalins can arise by proteolysis of larger precursor molecules. The relationship is nicely illustrated (Fig. 5) between Proopiocortin, a molecule of approximately 31,000 daltons, the intermediate sized adrenocorticotrophic hormone (ACTH), the melanophore stimulating hormones (Alpha-MSH and Gamma-MSH) and the opioid peptide B-endorphin. From the pioneer work of Bergmann and Fruton (1969), the group-selective enzymatic cleavage of linkages between amino acids by proteases and esterases accounts for their differing substrate specificities. For example, although chymotrypsin and trypsin are both serine proteases, sharing high reactivity toward organophosphate anticholinesterases, their substrate specificities are quite distinct. Trypsin shows specificity for peptides containing the basic aminoacid arginine to which is attached an aromatic aminoacid on an alpha-amino-group. In contrast, chymotrypsin prefers the opposite arrangement, i.e., a peptide containing a C-terminal aminoacid to which an aliphatic aminoacid forms the α-amino substituent. The specificity is not always restricted to "aromatic" bonds. Leucyl-, methionyl-, asparaginyl- and glutamyl-bonds are also hydrolyzed at high rates. The selective cleavage of pro-opiocortin by trypsin-like, chymotrypsin-like or exo- and endopeptidase activities in brain leads to the production of peptides with potent activity as neuromodulators or perhaps neurotransmitters. In commenting on the observation that brain contains its own morphine-like chemicals, Iversen presents the known structures of some of these peptides which are opioid-like in their actions, and includes yet another peptide α-neoendorphin with unique selective receptor binding activity (Fig. 6). He observes that the shorter chain peptides are metabolized more rapidly than their longer chain counterparts. The potency of the shorter chain dynorphin$_{1-8}$, was markedly increased with the addition of peptidase inhibitors. Though a common amino-acid sequence may be shared by biologically active substances of varying length, structures may not originate from a common precursor. It is instructive, however, to consider such relationships since it suggests that a large number of active fragments can derive from a relatively small number of differing macromolecules (Fig. 5). It can be seen that the aminoacid sequence (1-39) and beta-lipotropin (1-91) share a commonality with the macromolecule, pro-opiocortin. Alpha-melanophore stimulating hormone (α-MSH), vasoactive intestinal peptide (VIP), beta-MSH and beta-endorphin are included for they too possess partial sequences as well. The sequence of amino acids 61 through 65 is that of an Enkephalin, and B-endorphin has the sequence found from amino-acid 61 through 91. Methionine enkephalin (61-65) is included although there is evidence that the enkephalins and the newer opioid peptides, Dynorphin and α-neoendorphin (see Fig. 6) are formed from different precursors. It is though however, that the macromolecules

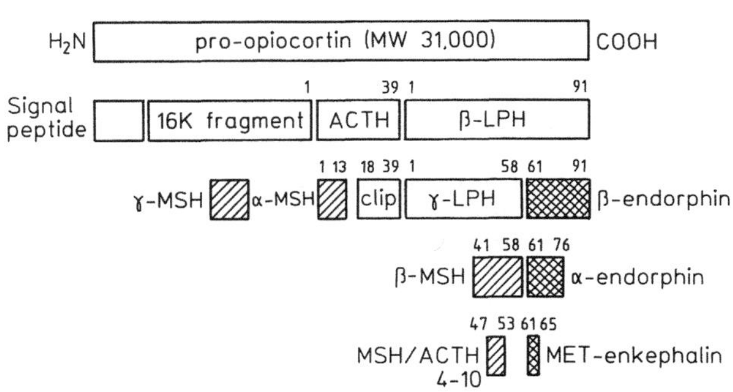

Fig. 5. The sequence of amino acids common to several biologically active peptides and where the same sequence is found in the protein pro-opiocortin. Although all peptides do not arise from pro-opiocortin their sequences do suggest a common origin

[1] Leu-enkephalin	Tyr-Gly-Gly-Phen-Leu	Fig. 6
[2] Dynorphin$_{1-8}$	Tyr-Gly-Gly-Phe-Leu-Arg-Arg-Ile	
[3] Dynorphin$_{1-17}$	Tyr-Gly-Gly-Phe-Leu-Arg-Arg-Ile-Arg-Pro-Lys-Leu-Lys-Trp-Asp-Asn-Gln	
[4] α-Neoendorphin	Tyr-Gly-Gly-Phe-Leu-Arg-Lys-Tyr-Pro-Lys	

[1] Huges, J. et al. 1975; [2] Weber, E. et al. 1982;[3] Goldstein, A. et al. 1979; [4] Kangawa, K. et al. 1981

Fig. 6. Dynorphin-related Peptides illustrating peptide homology

precursors arise from DNA of similar ancestral genotypes. An intensive study is currently underway in several prestigious laboratories to isolate and characterize those mRNA's responsible for the formation of pro-opiocortin and any other precursor molecules.

With the possible exception of TRH, peptides with important endocrine and/or neural function arise from partial degradation from high molecular weight precursors. To achieve the faithful reproduction of peptides having such defined sequence and chain length would require degrading enzymes with considerable group specificity. Such specificity, sharing common features at the active center, is an attribute of the family of serine proteases (Schachter 1980). Although additional evidence is needed to prove the existence in brain of such enzymes, including their isolation and characterization, indirect evidence supports their existence. What is more important, in the present context, is that the various peptides which have been obtained from brain are believed to arise from the selective proteolytic cleavage of a larger molecular weight substance. Such a hypothetical "master protein" was visualized, with tongue in cheek, by Professor Hokfelt, to consist of the sequence shown in Fig. 7. Included are peptides originally considered to occur only outside the nervous system, but now proven to be present, such as Cholecystokinin (CCK), Vasoactive Intestinal Peptide (VIP), Angiotension (ANG), Substance P (SP) and also those within the brain but thought of as performing only endocrine function e.g., ACTH, Somatostatin (SOM), Thyrotropin Releasing Hormone (TRH), Oxytocin (OXY) and Luteinizing Hormone Releasing factor (LHRH). The list of peptides, now known to be present in brain and nervous tissue, with marked excition or inhibition potentiating activity continues to expand. Agents which interfere with their formation or degradation would be expected to exert a significant effect on brain and nerve function.

The brain peptides, B-endorphin, leu-enkephhalin (Try-Gly-Gly-Phen-Leu) and met-enkephhalin (Try-Gly-Gly-Phen-Met) bind to opiate receptors and act as neurotransmitters or neuromodulators. The pharmacologic properties of B-endorphin are illustrative of the diverse actions of peptides (Fig. 8). In addition to the well known morphine-like analgesic action, beta-endorphin also exerts profound behavioral effects. In humans, ad-

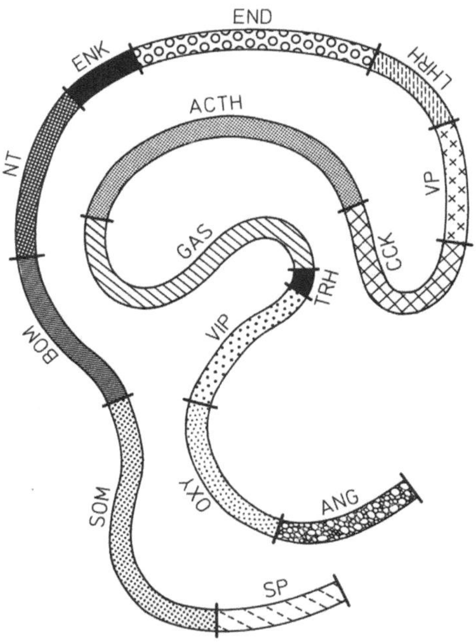

Fig. 7. A hypothetical "master protein' from which biologically active peptides could owe their evolutionary origin (from a lecture titled "Neuroanatomy for the Neurochemist", March 14, 1979, at the 10th Annual Meeting, American Society for Neurochemistry, Charleston, S.C.)

β-Endorphin
Analgesia
Physical Dependence
Tolerance
Cross-Tolerance
Euphoria
Resp. Depression
Constipation
Hormonal Effects
Behavioral Effects

Fig. 8. A partial list of the pharmacological properties of β-endorphin shown to emphasize the central effects of this peptide

ministration of the reversible anticholinesterase, physostigmine, produces significant elevations in plasma B-endorphin immunoreactivity and alterations in mood, cognition, and behavior (Risch et al. 1980). The results suggest to the authors that a "cholinergically mediated B-endorphin pathway may be involved in the observed affective changes". In a subsequent paper the authors offer evidence that a muscarinic site is involved in B-endorphin release (Risch et al. 1981). These and other findings in man and animal studies may be significant in organophosphate poisoning in light of Grob's description of the CNS effects produced. In man, cases of nonlethal exposure to organophosphate anticholinesterases, result in a number of CNS effects not exclusively explained by acetylcholine intoxication. According to Grob (1963), the earliest manifestations symptomatic of organophosphate poisoning usually include tension, anxiety, restlessness, emotional lability and giddiness. There may be insomnia, excessive dreaming, and, occasionally, nightmares.

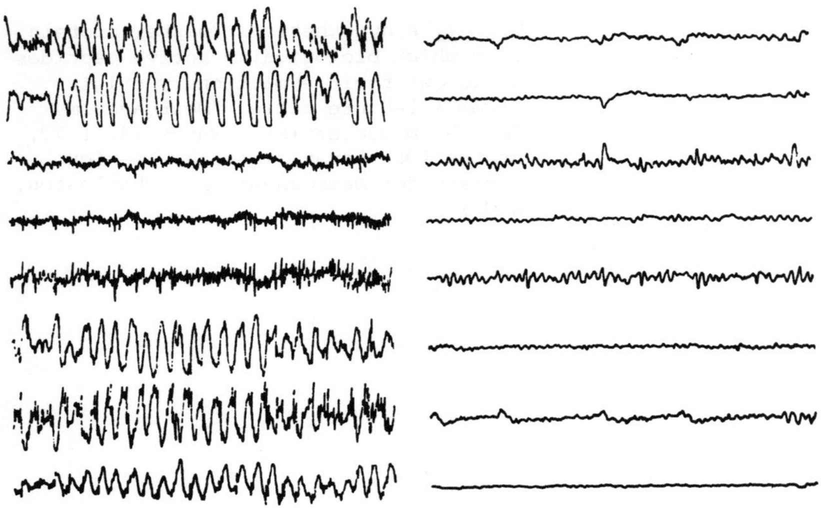

Fig. 9. Effect of marked exposure to Sarin on the electroencephalogram. Left, following exposure to Sarin, showing high-voltage slow waves in temporofrontal and temporocentral leads. Right, two weeks later, showing return to normal. Bipolar leads: left prefrontal to left anterior temporal (top), left anterior temporal to left low central, left low central to left low occiput, left occiput to right occiput, right occiput to right low central, right low central to right anterior temporal, right anterior temporal to right prefrontal, right prefrontal to left prefrontal (bottom). (From Grob 1963)

If the exposure is greater, there follows headache, tremor, difficulty in concentrating, and impairment of short-term memory. In some individuals there is apathy, withdrawal and depression. Rubin and Goldberg (1957a) and Rubin et al. (1957b) reported elevation of dark adaption and a rise in threshold of scotopic vision, *all* of a probable central origin. With the appearance of moderate symptoms, there occur abnormalities of the electroencephalographic patterns similar to those seen in temporal lobe epilepsy. When exposure is more intense generalized convulsions may ensue.

Grob (1963) described the influence of Sarin (methyl-isopropyl-phosphonofluoridate) exposure on those EEG patterns (Fig. 9) thus: "With the appearance of moderate toxic symptoms there occur abnormalities of the electronencephalogram, characterized by irregularities in rhythm, variation and intermittent bursts of abnormally slow waves of elevated voltage similar to those seen in patients with epilepsy. These abnormal waves become more marked after one or more minutes of hyperventilation, which if prolonged, may occasionally precipitate a generalized convulsion". These EEG changes may appear normal two weeks later or persist for several weeks; other CNS symptoms persisting for several months have been the complaint of some individuals. It is interesting in this regard that the intracerebroventricular administration of B-endorphin results in an initial epileptiform EEG response followed by subsequent onset of slow wave cortical

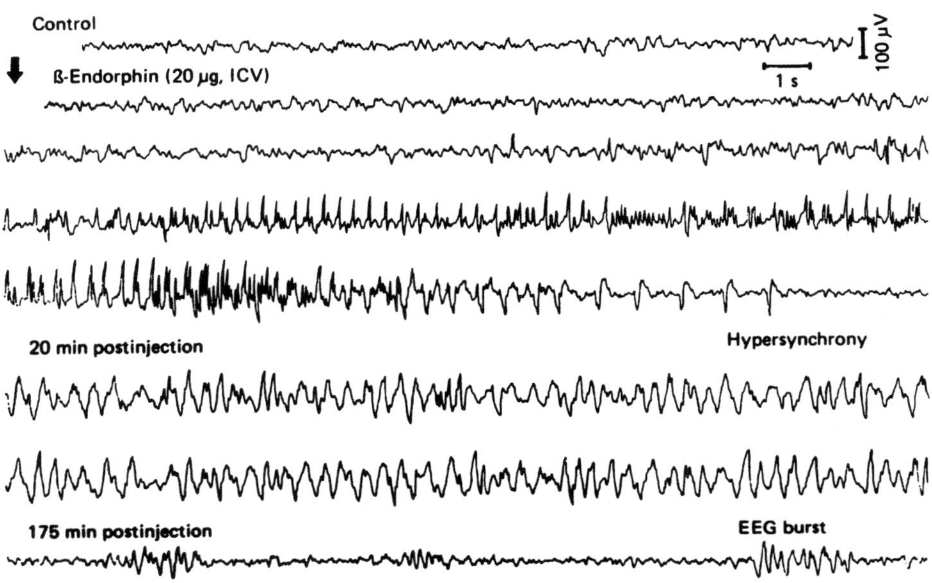

Fig. 10. Control and β-endorphin-induced patterns of EEG activity were measured in conscious rats from right frontal-right parietal cortical leads. Note initial epileptiform spike activity followed by the onset of high amplitude, slow wave hypersynchrony

EEG hypersynchrony, associated with increased voltage output according to Moreton et al. (1978); Tortella et al. (1978) (Fig. 10). In similar studies, the enkephalins, as well as beta-endorphin were shown to produce EEG seizure activity when injected intraventricularly in rats (Snead et al. 1980). The seizures are overcome by Tridione, Ethosuximide and Valproate, drugs effective in treating petite mal; Phenobarbital and Phenytoin were ineffectual. The seizure pattern was also antagonized by Naloxone, but at relatively high doses (0.1-10 mg/kg). The presence of specific exo- and endopeptidases is known to cause a breakdown of enkephalins when injected and probably accounts for their relatively short duration of action. The administration of DFP to inhibit their hydrolysis prolongs their activity.

The description of symptoms in humans by Grob (1963) following varying degrees of exposure to organophosphate anticholinesterases were difficult at the time to explain. Some, but not all of the effects seemed to be due to acetylcholine intoxication but symptoms persisted even in the presence of atropine. This by itself is not too surprising since atropine does not affect nicotinic receptors. Also it has long been recognized that stimulation of some neural elements results in response not blocked by either cholinergic or adrenergic antagonists. For example, the observation by Langley and Anderson (1895) that sacral stimulation of the urinary bladder was resistant to atropine. This early finding foreshadowed the description of a biologically active principle in tissue designated as Substance

Substance P

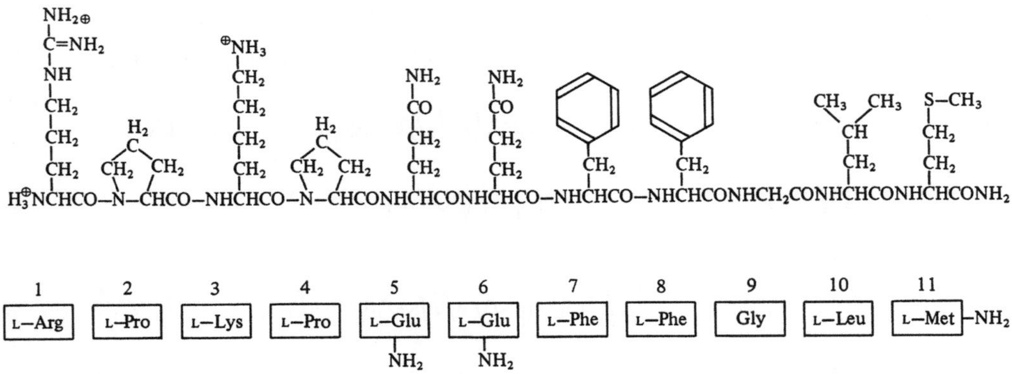

Fig. 11. The structure of substance P is shown to illustrate the amino acid sequence. Of special interest is linkage of L-pro to L-glu which appears to be a site attacked by prolidases

P by von Euler and Gaddum (1931) which accounted for Langley's observation. The existence of Substance P in brain and gut preceded by several decades, the recognition of other active peptides in the central and peripheral nervous systems. Substance P (Fig. 11) was finally isolated in pure form and its structure established by Chang and Leeman (1970) and Chang et al. (1971). It is a peptide containing 11 L-aminoacids whose sequence was established by enzymatic cleavage with α-chymotrypsin and carboxypeptidase-A. The occurrence in brain and nervous tissue of chymotryptic or similar enzyme activities could terminate the actions of Substance P.

As indicated previously, just as there are proteases and exo- and endopeptidases present in tissue responsible for the formation of the enkephalins, several enkephalin degrading enzymes are known to be present in brain. Their presence accounts for the transitory nature of enkephalin effects. The principal action in brain (and serum) by aminopeptidase(s) is cleavage of enkephalins at the N-terminal Tyr-Gly bond. In addition to the cytoplasmic enzyme there is a membrane bound carboxypeptidase which hydrolyzes the Gly-Phe bond to release either Phe-Leu from leu-enkephalin or Phe-Met from met-enkephalin (Benuck and Marks 1980, and Schnebli et al. 1979). There appears to be a number of additional peptidases whose specific role in the regulation of peptide activity in brain remains to be established, according to these authors.

Many biologically active peptides contain a proline residue (Substance P, neurotensin, TRH and alpha-neo endorphin, a precursor of leu-enkephalin). The modulatory role of these peptides may in part be regulated by the action of enzyme(s) hydrolyzing proline peptide bonds. Isolation of a prolyl-endopeptidase has been described from rat brain, which hydrolyzes Substance P specifically at the Pro-Glu bond (Kato et al. 1980). It is an enzyme which is completely inhibited by diisopropylphosphofluoridate (DFP) with an $IC_{50} = 10^{-6}M$. Similar prolyl-enzyme activities have been described in rabbit and human brain.

Interest in Substance P as a neurotransmitter or as a neuromodulator at certain sites in the central and peripheral nervous system is based on many findings. There is evidence that Substance P depresses the nicotinic excitation of Renshaw cells following iontophoresis of acetylcholine in spinal cord (Krnjevic and Leikic 1977). In isolated chromaffin cells, Substance P also acts to modulate the nicotinic acetylcholine response. This lends further support for the role of Substance P as an inhibitory neuromodulator at nicotinic receptor sites (Livett et al. 1979). With respect to the modulating influence of peptides, the work of Malthe-Sorensen et al. (1979) in Fonnum's laboratory is especially interesting. The local injection of Substance P into the septum of the rat specifically decreased acetylcholine turnover in the hippocampus. The authors suggest a certain regional distribution of this peptide whose action is to modulate the actions of primary transmitters.

A recent paper by Chubb et al. (1980), raises an interesting point concerning the possible hydrolysis of Substance P by acetylcholinesterase (AChE). The well established presence of AChE in the dorsal horn of the spinal cord has always been puzzling as this structure does not contain acetylcholine nor the ACh synthesizing enzyme, choline acetyltransferase (Giacobini 1956; Hebb and Selner 1956). Substance P is released at nerve endings from sensory afferent stimulation. Highly purified preparations of acetylcholinesterase were obtained by affinity chromatography from fetal calf serum and from electric eel electroplax. Both esterases were shown to release the terminal methionine and phenylalanine groups from Substance P. DFP and ACh inhibited the degradation of Substance P although the concentrations required were quite high (Fig. 12). DFP inhibition was maximal at $10^{-3}M$ of the compounds tested. No other substance prevented the degradation of Substance P including edro-

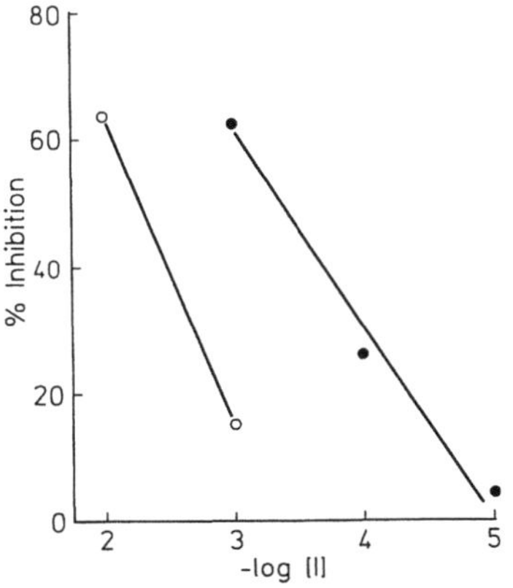

Fig. 12. The inhibition of substance P hydrolysis by acetylcholine and diisopropylphosphorofluoridate. The enzyme was pre-incubated for 10 min at 37°C with DFP (●); there was no pre-incubation with ACh (o)

phonium, BW 284C51, and eserine. These results could likely be explained by the presence of a second protein which accompanies AChE throughout the purification, although the authors tend to discount the possibility. Further work is needed in theirs and other laboratories to confirm the co-identity of AChE-activity with Substance P proteases. This is especially true since the inhibition by DFP of the AChE activity occurs at concentrations several orders of magnitude lower than that found in Chubb's study. Nevertheless, the presence of peptidase activity in close relationship to cholinesterase in areas such as the dorsal spinal roots and its inhibitability by organophosphates re-emphasizes the possibility that the anticholinesterases may exert actions toxic to the nervous system in unanticipated ways. Fresh insights into possible actions of organophosphates can be gained by reviewing recent developments in the neuropeptide field. Such approaches may prove rewarding in the study of the "non-cholinesterase actions" of the organophosphate anticholinesterases.

C. Conclusion

The presence in the nervous tissue of extremely potent peptides arises from the enzymatic degradation of larger molecular weight proteins. These smaller units are thought to exert modulatory actions on known transmitters. Their short half-life when injected is due to the degradative action of specific exo- and endopeptidases. It is suggested from the meagre evidence available that organophosphates may adversely affect the processes of formation and/or degradation to account for many of the adverse effects of organophosphate anticholinesterases which cannot be attributed to acetylcholine accumulation.

It is known that the opioid peptides B-endorphin and the enkephalins, produce epileptiform EEG patterns when administered intracerebrally in rats. The characteristic changes from normal patterns are thought to be of limbic origin according to Hendriksen et al. (1978). The alterations, which are not associated with overt convulsions, are sensitive to naloxone antagonism and, except in the case of D-ala^{2}-met-enkephalinamide a synthetic homolog of met-enkephalin with greater stability to peptidase action, tolerance has not been observed. These observations would suggest that lack of stability to "enkephalinase" actions may be an important attribute of the naturally formed opioids and explain in part their short duration of action. Inhibition of such peptidases could lead to a persistance of action and account for the altered EEG patterns observed in organophosphate poisoning. An apparent paradoxical effect of the opioid peptides is that the excitatory actions in producing EEG seizures contrasts with the usual inhibitory actions of the opiates and opioid peptides in most brain areas (Nicoll et al. 1977). The authors suggest that opiate-evoked epileptiform activity arises from a *disinhibition* of inhibitory interneurons in the proximity of the hippocampus, resulting in excitation of hippocampal pyramidal neurons leading to seizure discharge. The naturally occurring enkephalins and B-endorphins produce EEG-seizure activity when injected intraventricularly in rats, according to a recent paper by Snead et al. (1980). Consistent with the findings of Tortella et al. (1979), seizure activity was antagonized by naloxone but at relatively high doses (0.1-10 mg/kg). The seizures were controlled more effectively by Tridione, Ethosuximide and Valproate, drugs effective in treating absence seizures (petite mal); Phenobarbital and Phenytoin were ineffectual. The presence of specific exo- and endopeptidases causes a breakdown of enkephalins and probably accounts for their relatively short duration of action when injected. The administration of DFP to inhibit their hydrolysis prolongs their activity.

The possibility that many of the actions encountered in organophosphate poisonings, accidental or otherwise, may be due to substances other than acetylcholine is important to an understanding of the processes involved. This may be important because it suggests therapeutic measures which may be taken, in addition to those currently employed as standard treatment for non-lethal exposure to anticholinesterase agents commonly used in agriculture.

It can be seen that in rat, only about one-third of the TRH content is in the hypothalamus with most being in extrahypothalamic areas notably in the preoptic area and septal region thalamus amygdala and brainstem. There is a rough correlation between TRH-distribution (Table 1) and binding sites (Table 2) according to Burt and Snyder (1975).

The effect of TRH in the central nervous system is to cause arousal and antagonize barbiturate anesthesia. The pharmacologic effects of microinjected TRH however seem to be dependent on the arousal state. When administered into the dorsal hippocampus of the awake animal there was a dose-dependent behavioral quieting of the animal and reduction in metabolism and electromyographic activity. In contrast, when administered during slow-wave sleep there was an increase in electromyographic activity and an increase in electroencephalographic desynchronization, similar to that seen in barbiturate anesthesia.

Table 1. Distribution of TRH in the CNS

Brain area	Area weight (Mg)	TRH content (Ng)	% TRH
Hypothalamus	32.6 ± 1.5	4.1 ± 0.2	31.2 ± 2.3
Forebrain	399.0 ± 8.9	3.5 ± 0.3	25.6 ± 2.1
Brainstem	185.0 ± 5.5	2.1 ± 0.1	16.9 ± 1.8
Posterior diencephalon	213.3 ± 8.7	1.9 ± 0.2	13.8 ± 1.7
Posterior Cortex	622.8 ± 11.5	1.3 ± 0.1	10.6 ± 0.7
Cerebellum	244.0 ± 5.1	0.26 ± 0.03	2.1 ± 0.3

Table 2. Regional distribution of 3H TRH binding in rat brain[a]

Region (N)	High-affinity binding		Low-affinity binding	
	No. of sites		No. of sites	
	$K_{D(NM)}$	(fmol/mg protein)	$K_{D(NM)}$	(pmol/mg protein)
Cerebral Cortex (7)	47 ± 5	110 ± 13	5 ± 1	30 ± 2
Hypothalamus (2)	52	190	9	46
Hypocampus (1)	36	150	3	39
Midbrain (1)	32	100	6	34
Corpus Striatum (1)	38	60	7	35
Pons-medulla (1)	27	40	2	7
Cerebellum (2)	N.D.	N.D.	0.9	7

[a]Adopted from Burt and Snyder (1975).

References

Bargmann, W., Lindner, E., Andres, K.H.: Über Synapsen an Endokrinen Epithelzellen und die Definition sekretorischer Neurone. Untersuchungen am Zwischenlappen der Katzenhypophyse. Z. Zellforsch. 77, 282-298 (1967)

Benuck, M., Marks, N.: Characterization of a distinct membrane bound dipeptidyl carboxypeptidase inactivating enkephalin in brain. BBRC 95, 822-828 (1980)

Bergmann, M.: A Classification of Proteolytic Enzymes. In: Advances in Enzymology (eds. F.F. Nord, C.H. Werkman), Vol. II, pp. 49-67. New York-London: Interscience Publishers 1942

Bergmann, M., Fruton, J.S.: The specificity of proteinases. In: Advances in Enzymology (eds. F.F. Nord, C.H. Werkman), Vol. I, pp. 63-96. New York-London: Interscience Publishers 1941

Bohler, J., Enzmann, F., Folkers, K., Bowers, C.Y., Schally, A.V.: The identity of chemical and hormonal properties of the thryrotropin releasing hormone and pyroglutamyl-histidyl-prolineamide. Biochem. Biophys. Res. Commun. 37, 705 (1969)

Breese, G.R., Cott, J.M., Cooper, B.R., Prange, A.P., Jr., Lipton, M.A.: Effects of thyrotropin releasing hormone (TRH) on the actions of pentobarbital and other centrally acting drugs. J. Pharm. Exp. Ther. 193, 11-22 (1975)

Breese, G.R., Mueller, R.A., Mailman, R.B., Frye, G.D.: Effects of TRH on central nervous system function. In: The Role of Peptides and Amino Acids as Neurotransmitters, pp. 99-116. New York: Liss 1981

Burgus, R., Dunn, T.F., Desiderio, D., Ward, D.N., Vale, W., Guillemin, R.: Characterization of ovine hypothalamic hypophysiotropic TSH releasing factor. Nature (Lond.) 226, 321-325 (1970)

Burt, D.R., Snyder, S.H.: Thyrotropin-releasing hormone (TRH): Apparent receptor binding in rat brain membranes. Brain Res. 92, 309-328 (1975)

Chang, M.M., Leeman, S.E.: Isolation of a sialogogic peptide from bovine hypothalamic tissue and its characterization as substance P. J. Biol. Chem. 245, 4787-4790 (1970)

Chang, M.M., Leeman, S.E., Niall, H.D.: Aminoacid sequence of substance P. Nature 232, 86-87 (1971)

Chretien, M., Crine, P., Seidah, N.G., Scherrer, H.: Biosynthesis of beta-lipoprotein and beta-endorphin: A model for other neuropeptides. In: Brain Peptides: A New Endocrinology (eds. A.M. Gotto, E.J. Peck, A.E. Boyd), pp. 161-172. Amsterdam-New York: Elsevier/North-Holland 1979

Chretien, M., Li, C.H.: Isolation, purification and characterization of γ-lipotropic hormone from sheep pituitary glands. Can. J. Biochem. 45, 1163-1174 (1967)

Chubb, I.W., Hodgson, A.J., White, G.H.: Acetylcholinesterase hydrolyzes substance P. Neuroscience 5, 2065-2072 (1980)

Cohen, J.A., Oosterbaan, R.A.: The active site of acetylcholinesterase and related esterases and its reactivity towards substrates and inhibitors. In: Handbuch der Experimentellen Pharmakologie (ed. G.B. Koelle), Vol. 15, pp. 351-352. Berlin-Heidelberg-New York: Springer 1963

Du Bois, K.P.: Toxicological evaluation of the anticholinesterase agents. In: Handbuch der Experimentellen Pharmakologie (ed. G.B. Koelle), Vol. 5, p. 54. Berlin-Heidelberg-New York: Springer 1963

Emson, P.C.: Peptides as neurotransmitter candidates in the mammalian CNS. Proc. Neurobiol. 13, 61 (1979)

Giacobini, E.: Quantiative determination of cholinesterase in individual spinal ganglion cells. Acta. Physiol. Scand. 45, 238-254 (1956)

Goldstein, A., Tachibana, S., Lowney, L.I., Hunkapillar, M., Hood, L.: Dynorphin-(1-13), an extraordinarily potent opioid peptide. Proc. Natl. Acad. Sci. USA 76, 6666-6670 (1979)

Grob, D.: in: Handbuch der Experimentellen Pharmakologie (ed. G. Koelle), Vol. 15, pp. 989-1027. Berlin-Heidelberg-New York: Springer 1963

Habener, J.F., Chang, J.T., Potts, J.R., Jr.: Enzymic processing of proparathyroid hormone by cell-free extracts of parathyroid glands. Biochem. 16, 3910-3917 (1977)

Hebb, C.D., Selner, A.: Cholineacetylase in the central nervous system of man and other mammals. J. Physiol. (London) 134, 718-728 (1956)

Hendriksen, S.J., Bloom, F.E., McCoy, F., Lang, N., Guillemin, R.: B-endorphin induces nonconvulsive limbic seizures. Proc. Natl. Acad. Sci. USA 75, 5221-5225 (1978)

Hokfelt, T., Fuxe, K., Johansson, O., Jeffcoate, S., White, N.: Distribution of thyrotropin-releasing hormone (TRH) in the central nervous system as revealed with immunohistochemistry. Europ. J. Pharmacol. 34, 389-392 (1975)

Hokfelt, T.: Neuroanatomy for the Neurochemist. The First Annual Basic Neurochemistry Lecture. Tenth Annual Meeting of the American Society for Neurochemistry. Charleston, SC 1979

Holmstedt, B.: Structure-activity relationships of the organophosphorus anticholinesterase agents. In: Handbuch der Experimentellen Pharmakologie: Cholinesterase and Anticholinesterases Agents, (sub-ed. G.B. Koelle), Vol. 15, pp. 428-485. Berlin-Göttingen-Heidelberg: Springer 1963

Hughes, J., Smith, T.W., Kosterlitz, H.W.: Identification of two related pentapeptides from the brain with potent opiate agonist activity. Nature 258, 577 (1975)

Jackson, I.M.D., Reichlin, S.: Distribution and biosynthesis of TRH in the nervous system. In: CNS Effects of Hypothalamic Hormones and Other Peptides (ed. Collu et al.), pp. 3-54. New York: Raven Press: 1979

Jandorf, B.J., Wagner-Jauregg, T., O'Neill, J.J., Stolberg, M.A.: The reaction of phosphorus containing enzyme inactivators with phenols and polyphenols. J. Am. Chem. Soc. 74, 1521-1523 (1952)

Jandorf, B.J., Michel, H.O., Schaffer, N.K., Egan, R., Summerson, W.H.: The mechanism of reaction between esterases and phosphorus-containing antiesterases. In: The Physical Chemistry of Enzymes. Disc. Faraday Soc., No. 20, pp. 134-142. Aberdeen: University Press 1955

Jansen, E.F., Nutting, M.D.F., Ball, A.K.: Mode of inhibition of chymotrypsin by DFP. J. Biochem. 179, 201-204 (1949)
Johansson, O., Hokfelt, T.: Immunohistochemical distribution of thyrotropin releasing hormone, somatostatin and enkephalin with special reference to the hypothalamus. In: Brain and pituitary peptides, pp. 202-242. Ferring Symp., Munich 1979. Basel: Karger 1980
Kalivas, P.W., Halperin, L.M., Horita, A.: Synchronization of hippocampal and cortical electroencephalogram by thyrotropin releasing hormone. Exp. Neurol. 69, 627-638 (1980)
Kangawa, K., Minamino, N., Chino, N., Sakakibara, S., Matsuo, H.: The complete amino acid sequence of alpha-neo-endorphin. Biochem. Res. Commun. 99, 871-878 (1981)
Kato, T., Nakano, T., Kojima, K., Nagatsu, T., Sakakibara, S.: Changes in prolyl endopeptidase during maturation of rat brain and hydrolysis of substance P by the purified enzyme. J. Neurochem. 35 (3), 527-535 (1980)
Koshland, D.E.: The active site and enzyme action. In: Advances in Enzymology (ed. F.F. Nord), Vol. 22, pp. 45-907. New York-London: Interscience Publishers 1960
Kossiakoff, A.A., Spencer, S.A.: Neutron diffraction identifies His^{57} as the catalytic base in trypsin. Nature 288, 414-416 (1980)
Kraut, J.: Serine Proteases: Structure and Mechanism of Catalysis. Annu. Rev. of Bicohem. 46, 331-358 (1977)
Krnjevic, K., Leikic, D.: Substance P selectively blocks excitation of renshaw cell by acetylcholine. Can. J. Physiol. Pharmacol. 55, 958-961 (1977)
Langley, J.N., Anderson, H.K.: The innervation of the pelvic and adjoining viscera. J. Physiol. (London) 19, 71-131 (1895)
Larsson, L.I., Rehfeld, J.F.: Localization and molecular heterogeneity of cholecystokinin in the central and peripheral nervous system. Brain Res. 165 (2), 201-218 (1979)
Livett, B.G., Kozousek, V., Mizobe, F., Dean, D.M.: Substance P inhibits nicotinic activation of chromaffin cells. Nature 278, 256-257 (1979)
Malthe-Sorenson, D., Wood, P.: Modulation of acetylcholine turnover rate by neuropeptides. Prog. Brain Res. 29, 486-487 (1979)
Malthe-Sorenson, D., Wood, Pl.L., Cheney, D.L., Costa, E.: Modulation of the turnover rate of acetylcholine in rat brain by intraventricular injections of thyrotropin releasing hormone, somatostatin, neurotensin and angiotension II. J. Neurochem. 31, 685-691 (1978)
Marks, N., Datta, R.K., Lajtha, A.: Partial resolution of brain arylamidases and aminopeptidases. J. Biol. Chem. 243, 282-2889 (1968)
Moreton, J.E., Tortella, F.C. Khazan, N.: EEG and behavioral effects of acute and chronic administration of opiate-like peptides in the rat. Dev. Neurosci. 4, 435-437 (1978)
Neurath, H., Walsh, K.A.: Role of proteolytic enzymes in biological regulation (a review). Proc. Natl. Acad. Sci. USA 73(11), 3825-3832 (1979)
Nicoll, R.A., Siggins, G.R., Long, N., Bloom, F.E., Guillemin, R.: Neuronal actions of endorphins and enkephalins among brain regions: A comparative microiontophoretic study. Proc. Natl. Acad. Sci. USA 74, 2584-2588 (1977)
Oosterbaan, R.A., Jansz, H.S., Cohen, J.A.: The chemical structure of the reactive group of esterases. Biochim. Biophys. Acta 20, 402-403 (1958)
Oosterbaan, R.A., Kunst, P., Van Rotterdam, J., Cohen, J.A.: The reaction of chymotrypsin and diisopropylphosphorofluoridate. Isolation and analysis of diisopropylphosphoryl peptides. Biochim. Biophys. Acta 27, 549-555 (1958); The Reaction of chymotrypsin and Diisopropylphosphorofluoridate. The structure of two DP-substitutes peptides from chymotypsin. Biochim. Biophys. Acta 27, 556-563 (1958)

Osbahr, A.J., Nemeroff, C.B., Lullinger, D., Mason, G.A., Prange, A.J., Jr.: Neurotensin-induced antinociception in mice. J. Pharmacol. Exp. Ther. 217, 645-651 (1981)
Phillis, J.W., Kirkpatrick, J.R., Said, S.I.: Vasoactive intestinal polypeptide excitation of central neurons. Can. J. Physiol. Pharmacol. 56, 337-340 (1978)
Phillis, J.W.: Responses of central neurons to neuronally localized peptides. In: Advances in Pharmacological Research and Practice Aminergic and Peptidergic Receptors (eds. E.S. Vizi, M. Wollemann), Vol. VII, PP. 151-168. Oxford: Pergamon 1979
Possible Long-Term Health Effects of Short-Term Exposure to Chemical Agents, Vol. I. Anticholinesterases and Anticholinergics, Comm. on Toxicol., Board on Toxicol. and Environ. Health Hazards, Assembly of Life Science, National Research Council, June 1982
Risch, S.C., Cohen, R.M., Janowsky, D.S., Kalin, N.H., Murphy, D.L.: Mood and behavioral effects of physostigmine on humans are accompanied by elevations in plasma B-endorphin and cortisol. Science 209, 2545-2546 (1980)
Risch, S.C., Kalin, N.H., Cohen, R.M., Weker, J.L., Insel, T.R., Cohen, M.L., Murphy, D.L.: Muscarinic cholinergic influences on ACTH and B-endorphin release mechanisms in human subjects. Peptides 2(1), 95-97 (1981)
Rubin, L.S., Goldberg, M.N.: Effect of sarin on dark adaptation in man: threshold changes. J. Appl. Physiol. II(3), 439-444 (1957a)
Rubin, L.S., Krop, S., Goldberg, M.N.: Effect of sarin on dark adaptation in man: mechanism of action. J. Appl. Physiol. III(3), 445-449 (1957b)
Ryder, S.W., Eng, J., Strauss, E., Yalow, R.S.: Extraction and immunochemical characterization of cholecystokinin-like peptides from pig and rat brain. Proc. Natl. Acad. Sci. USA 78, 3892-3896 (1981)
Schachter, M.: Kallikreins (Kininogenases) - A group of serine proteases with bioregulatory actions. Pharmacol. Rev. 31(1), 1-17 (1980)
Schaffer, N.K., Engle, R.R., Simet, L., Drisko, R.U.: Phosphopeptides from chymotrypsin and trypsin after inactivation by ^{32}P labeled DFP and sarin. Fed. Proc. 15, 347; J. Biochem. 230, 185-192 (1958)
Schnebli, H.P., Phillipps, M.A., Barclay, R.K.: Isolation and characterization of an enkephalin-degrading aminopeptidase from rat brain. Biochim. Biophys. Acta 569, 89-98 (1979)
Snead, O., III, Bearden, L.J.: Anticonvulsants specific for petit mal antagonize epileptogenic effect of leucine enkephalin. Science 210, 1031-1033 (1980)
Stanton, T.L., Beckman, A.L., Winokur, A.: Thyrotropin releasing hormone effects in the central nervous system: dependence on arousal state. Science 214, 678-681 (1981)
Sternberger, L.A.: Immunocytochemistry. Englewood Cliffs: Prentice-Hall 1974
Tortella, F.C., Cowan, A., Adler, M.W.: Comparison of the anticonvulsant effects of opioid peptides and etorphine in rats after ICV administrant. Life Sci. 10, 1039-1045 (1981)
Wagner-Jauregg, T., O'Neill, J.J., Summerson, W.H.: The reaction of phosphorus-containing enzyme inhibitors with amines and amino acid derivatives. J. Am. Chem. Soc. 73, 5202-5206 (1951)
Weber, E., Evans, C.J., Barchas, J.D.: Predominance of the amino-terminal octapeptide fragment of dynorphin in rat brain regions. Nature 299, 77-79 (1982)
Wescoe, W.C., Greene, R.E., McNamara, R., Krop, S.: The influence of atropine and scopolamine on the central effects of DFP. J. Pharmacol. Exp. Ther. 92, 63-72 (1948)

Subject Index

MIX
Papier aus verantwortungsvollen Quellen
Paper from responsible sources
FSC® C105338

If you have any concerns about our products,
you can contact us on
ProductSafety@springernature.com

In case Publisher is established outside the EU,
the EU authorized representative is:
Springer Nature Customer Service Center GmbH
Europaplatz 3, 69115 Heidelberg, Germany

Printed by Libri Plureos GmbH
in Hamburg, Germany